Student Edition

Eureka Math
Algebra I
Modules 1, 2, & 3

Special thanks go to the Gordon A. Cain Center and to the Department of
Mathematics at Louisiana State University for their support in the development of
Eureka Math.

For a free *Eureka Math* Teacher Resource Pack, Parent Tip Sheets, and more please visit www.Eureka.tools

Published by Great Minds

Printed in the U.S.A.
This book may be purchased from the publisher at eureka-math.org
10 9 8 7 6 5 4 3 2 1
ISBN 978-1-63255-324-9

Lesson 1: Graphs of Piecewise Linear Functions

Classwork

Exploratory Challenge

Example 1

Here is an elevation-versus-time graph of a person's motion. Can we describe what the person might have been doing?

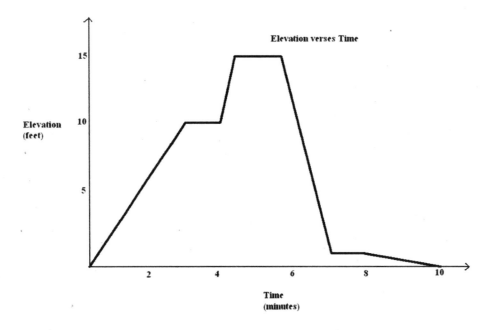

PIECEWISE-DEFINED LINEAR FUNCTION: Given non-overlapping intervals on the real number line, a *(real) piecewise linear function* is a function from the union of the intervals on the real number line that is defined by (possibly different) linear functions on each interval.

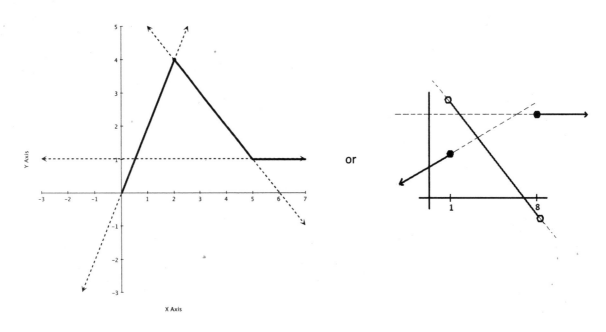

Lesson 1:   Graphs of Piecewise Linear Functions

EUREKA
MATH™

Problem Set

1. Watch the video, "Elevation vs. Time #3" (below)

   http://www.mrmeyer.com/graphingstories1/graphingstories3.mov. (This is the third video under "Download Options" at the site http://blog.mrmeyer.com/?p=213 called "Elevation vs. Time #3.")

   It shows a man climbing down a ladder that is 10 ft. high. At time 0 sec., his shoes are at 10 ft. above the floor, and at time 6 sec., his shoes are at 3 ft. From time 6 sec. to the 8.5 sec. mark, he drinks some water on the step 3 ft. off the ground. After drinking the water, he takes 1.5 sec. to descend to the ground, and then he walks into the kitchen. The video ends at the 15 sec. mark.

   a. Draw your own graph for this graphing story. Use straight line segments in your graph to model the elevation of the man over different time intervals. Label your x-axis and y-axis appropriately, and give a title for your graph.

   b. Your picture is an example of a graph of a piecewise linear function. Each linear function is defined over an interval of time, represented on the horizontal axis. List those time intervals.

   c. In your graph in part (a), what does a horizontal line segment represent in the graphing story?

   d. If you measured from the top of the man's head instead (he is 6.2 ft. tall), how would your graph change?

   e. Suppose the ladder descends into the basement of the apartment. The top of the ladder is at ground level (0 ft.) and the base of the ladder is 10 ft. below ground level. How would your graph change in observing the man following the same motion descending the ladder?

   f. What is his average rate of descent between time 0 sec. and time 6 sec.? What was his average rate of descent between time 8.5 sec. and time 10 sec.? Over which interval does he descend faster? Describe how your graph in part (a) can also be used to find the interval during which he is descending fastest.

2. Create an elevation-versus-time graphing story for the following graph:

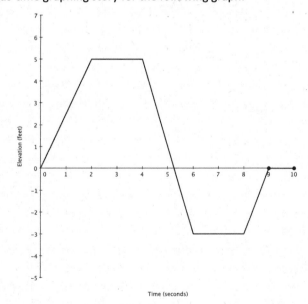

3. Draw an elevation-versus-time graphing story of your own, and then create a story for it.

This page intentionally left blank

Lesson 2: Graphs of Quadratic Functions

Classwork

Exploratory Challenge

Plot a graphical representation of change in elevation over time for the following graphing story. It is a video of a man jumping from 36 ft. above ground into 1 ft. of water.

http://www.youtube.com/watch?v=ZCFBC8aXz-g or http://youtu.be/ZCFBC8aXz-g (If neither link works, search for "OFFICIAL Professor Splash World Record Video!")

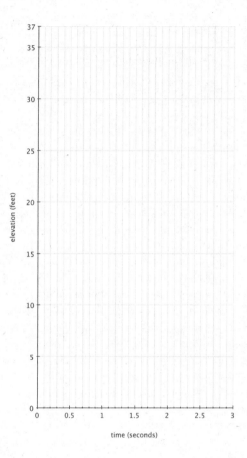

Example 2

The table below gives the area of a square with sides of whole number lengths. Have students plot the points in the table on a graph and draw the curve that goes through the points.

Side (cm)	0	1	2	3	4
Area (cm^2)	0	1	4	9	16

On the same graph, reflect the curve across the y-axis. This graph is an example of a graph of a quadratic function.

EUREKA
MATH™

Problem Set

1. Here is an elevation-versus-time graph of a ball rolling down a ramp. The first section of the graph is slightly curved.

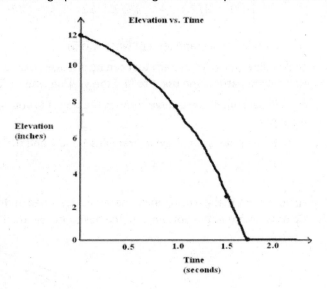

a. From the time of about 1.7 sec. onward, the graph is a flat horizontal line. If Ken puts his foot on the ball at time 2 sec. to stop the ball from rolling, how will this graph of elevation versus time change?

b. Estimate the number of inches of change in elevation of the ball from 0 sec. to 0.5 sec. Also estimate the change in elevation of the ball between 1.0 sec. and 1.5 sec.

c. At what point is the speed of the ball the fastest, near the top of the ramp at the beginning of its journey or near the bottom of the ramp? How does your answer to part (b) support what you say?

2. Watch the following graphing story:

Elevation vs. Time #4 [http://www.mrmeyer.com/graphingstories1/graphingstories4.mov. This is the second video under "Download Options" at the site http://blog.mrmeyer.com/?p=213 called "Elevation vs. Time #4."]

The video is of a man hopping up and down several times at three different heights (first, five medium-sized jumps immediately followed by three large jumps, a slight pause, and then 11 very quick small jumps).

a. What object in the video can be used to estimate the height of the man's jump? What is your estimate of the object's height?

b. Draw your own graph for this graphing story. Use parts of graphs of quadratic functions to model each of the man's hops. Label your x-axis and y-axis appropriately and give a title for your graph.

3. Use the table below to answer the following questions.

x	0	1	2	3	4	5	6
y	0	3/2	4	15/2	12		24

a. Plot the points (x, y) in this table on a graph (except when x is 5).

b. The y-values in the table follow a regular pattern that can be discovered by computing the differences of consecutive y-values. Find the pattern and use it to find the y-value when x is 5.

c. Plot the point you found in part (b). Draw a curve through the points in your graph. Does the graph go through the point you plotted?

d. How is this graph similar to the graphs you drew in Examples 1 and 2 and the Exploratory Challenge? Different?

4. A ramp is made in the shape of a right triangle using the dimensions described in the picture below. The ramp length is 10 ft. from the top of the ramp to the bottom, and the horizontal width of the ramp is 9.25 ft.

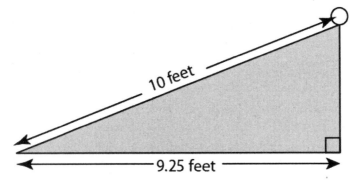

A ball is released at the top of the ramp and takes 1.6 sec. to roll from the top of the ramp to the bottom. Find each answer below to the nearest $0.1 \frac{\text{ft}}{\text{sec}}$.

a. Find the average speed of the ball over the 1.6 sec.

b. Find the average rate of horizontal change of the ball over the 1.6 sec.

c. Find the average rate of vertical change of the ball over the 1.6 sec.

d. What relationship do you think holds for the values of the three average speeds you found in parts (a), (b), and (c)? (Hint: Use the Pythagorean theorem.)

EUREKA
MATH™

Lesson 3: Graphs of Exponential Functions

Classwork

Example

Consider the story:

Darryl lives on the third floor of his apartment building. His bike is locked up outside on the ground floor. At 3:00 p.m., he leaves to go run errands, but as he is walking down the stairs, he realizes he forgot his wallet. He goes back up the stairs to get it and then leaves again. As he tries to unlock his bike, he realizes that he forgot his keys. One last time, he goes back up the stairs to get his keys. He then unlocks his bike, and he is on his way at 3:10 p.m.

Sketch a graph that depicts Darryl's change in elevation over time.

Exploratory Challenge

Watch the following graphing story:

https://www.youtube.com/watch?v=gEwzDydciWc

The video shows bacteria doubling every second.

a. Graph the number of bacteria versus time in seconds. Begin by counting the number of bacteria present at each second and plotting the appropriate points on the set of axes below. Consider how you might handle estimating these counts as the population of the bacteria grows.

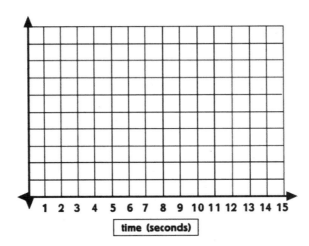

b. Graph the number of bacteria versus time in minutes.

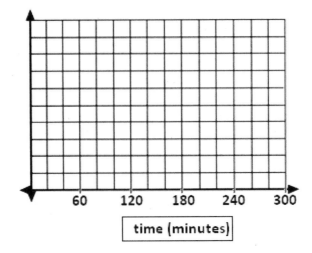

EUREKA
MATH™

c. Graph the number of bacteria versus time in hours (for the first five hours).

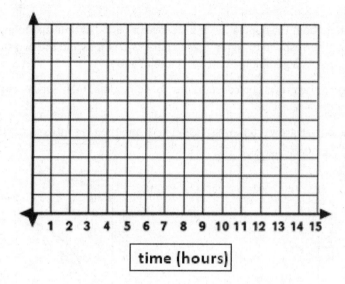

time (hours)

Problem Set

1. Below are three stories about the population of a city over a period of time and four population-versus-time graphs. Two of the stories each correspond to a graph. Match the two graphs and the two stories. Write stories for the other two graphs, and draw a graph that matches the third story.

   Story 1: The population size grows at a constant rate for some time, then doesn't change for a while, and then grows at a constant rate once again.

   Story 2: The population size grows somewhat fast at first, and then the rate of growth slows.

   Story 3: The population size declines to zero.

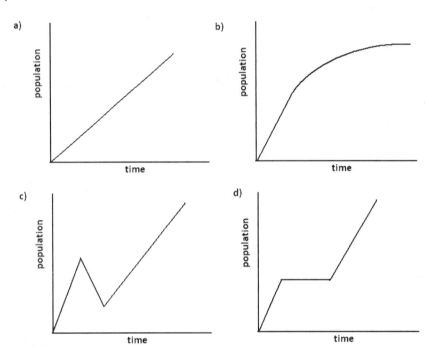

2. In the video, the narrator says:

   "Just one bacterium, dividing every 20 min., could produce nearly 5,000 billion billion bacteria in one day. That is 5,000,000,000,000,000,000,000 bacteria."

   This seems WAY too big. Could this be correct, or did she make a mistake? (Feel free to experiment with numbers using a calculator.)

3. *Bacillus cereus* is a soil-dwelling bacterium that sometimes causes food poisoning. Each cell divides to form two new cells every 30 min. If a culture starts out with exactly 100 bacterial cells, how many bacteria will be present after 3 hr.?

4. Create a story to match each graph below:

a)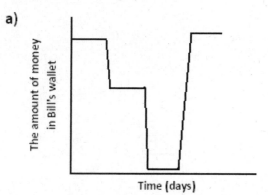

b)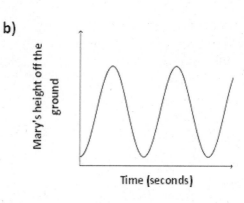

5. Consider the following story about skydiving:

Julie gets into an airplane and waits on the tarmac for 2 min. before it takes off. The airplane climbs to 10,000 ft. over the next 15 min. After 2 min. at that constant elevation, Julie jumps from the plane and free falls for 45 sec. until she reaches a height of 5,000 ft. Deploying her chute, she slowly glides back to Earth over the next 7 min. where she lands gently on the ground.

a. Draw an elevation-versus-time graph to represent Julie's elevation with respect to time.

b. According to your graph, describe the manner in which the plane climbed to its elevation of 10,000 ft.

c. What assumption(s) did you make about falling after she opened the parachute?

6. Draw a graph of the number of bacteria versus time for the following story: Dave is doing an experiment with a type of bacteria that he assumes divides in half exactly every 30 min. He begins at 8:00 a.m. with 10 bacteria in a Petri dish and waits for 3 hr. At 11:00 a.m., he decides this is too large of a sample and adds Chemical A to the dish, which kills half of the bacteria almost immediately. The remaining bacteria continue to grow in the same way. At noon, he adds Chemical B to observe its effects. After observing the bacteria for two more hours, he observes that Chemical B seems to have cut the growth rate in half.

7. Decide how to label the vertical axis so that you can graph the data set on the axes below. Graph the data set and draw a curve through the data points.

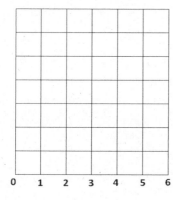

x	y
0	-1
1	-2
2	-4
3	-8
4	-16
5	-32
6	-64

This page intentionally left blank

Lesson 4: Analyzing Graphs—Water Usage During a Typical Day at School

Classwork

Example

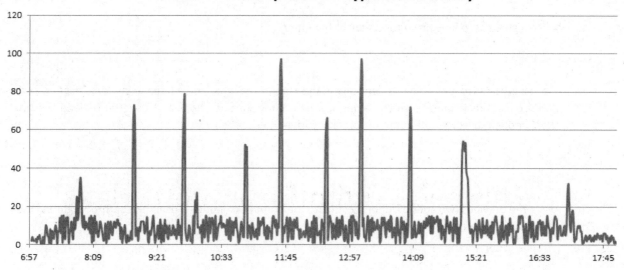

Water Consumption in a Typical School Day

Exercises 1–2

1. The bulk of water usage is due to the flushing of toilets. Each flush uses 2.5 gal. of water. Samson estimates that 2% of the school population uses the bathroom between 10:00 a.m. and 10:01 a.m. right before homeroom. What is a good estimate of the population of the school?

2. Samson then wonders this: If everyone at the school flushed a toilet at the same time, how much water would go down the drain (if the water pressure of the system allowed)? Are we able to find an answer for Samson?

Exercise 3: Estimation Exercise

3.

    a. Make a guess as to how many toilets are at the school.

    b. Make a guess as to how many students are in the school, and what percentage of students might be using the bathroom at break times between classes, just before the start of school, and just after the end of school. Are there enough toilets for the count of students wishing to use them?

    c. Using the previous two considerations, estimate the number of students using the bathroom during the peak minute of each break.

    d. Assuming each flush uses 2.5 gal. of water, estimate the amount of water being used during the peak minute of each break.

    e. What time of day do these breaks occur? (If the school schedule varies, consider today's schedule.)

    f. Draw a graph that could represent the water consumption in a typical school day of your school.

Problem Set

1. The following graph shows the temperature (in degrees Fahrenheit) of La Honda, CA in the months of August and September of 2012. Answer the questions following the graph.

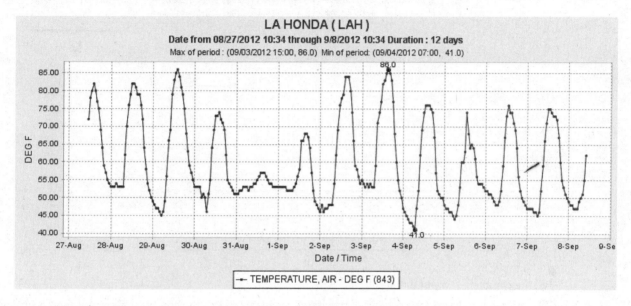

a. The graph seems to alternate between peaks and valleys. Explain why.

b. When do you think it should be the warmest during each day? Circle the peak of each day to determine if the graph matches your guess.

c. When do you think it should be the coldest during each day? Draw a dot at the lowest point of each day to determine if the graph matches your guess.

d. Does the graph do anything unexpected such as not following a pattern? What do you notice? Can you explain why it is happening?

2. The following graph shows the amount of precipitation (rain, snow, or hail) that accumulated over a period of time in La Honda, CA.

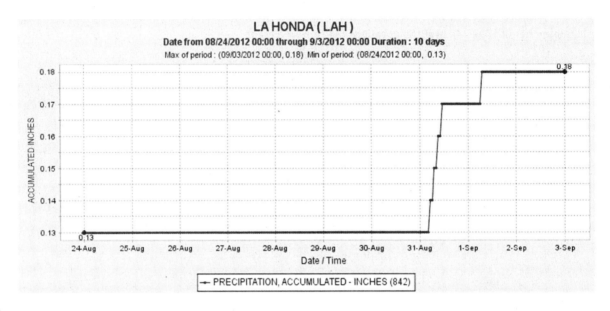

a. Tell the complete story of this graph.

b. The term *accumulate,* in the context of the graph, means to add up the amounts of precipitation over time. The graph starts on August 24. Why didn't the graph start at 0 in. instead of starting at 0.13 in.?

3. The following graph shows the solar radiation over a period of time in La Honda, CA. Solar radiation is the amount of the sun's rays that reach the earth's surface.

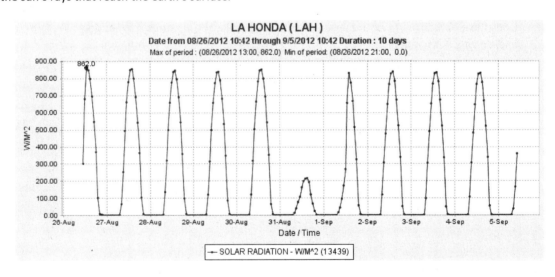

a. What happens in La Honda when the graph is flat?

b. What do you think is happening when the peaks are very low?

c. Looking at all three graphs, what do you conclude happened on August 31, 2012 in La Honda, CA?

4. The following graph shows the velocity (in centimeters per second) and turbidity of the Logan River in Queensland, Australia during a flood. Turbidity refers to the clarity of the water (higher turbidity means murkier water) and is related to the total amount of suspended solids, such as clay, silt, sand, and phytoplankton, present in the water.

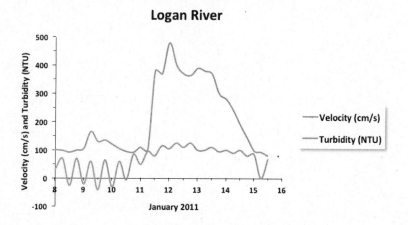

a. For recreation, Jill visited the river during the month of January and saw clean and beautiful water. On which day do you think she visited?

b. What do the negative velocities (below the grey line) that appear periodically at the beginning represent?

c. The behavior of the river seems to follow a normal pattern at the beginning and at the very end of the time period shown. Approximately when does the flood start? Describe its effects on velocity and turbidity.

This page intentionally left blank

Lesson 5: Two Graphing Stories

Classwork

Example 1

Consider the story:

Maya and Earl live at opposite ends of the hallway in their apartment building. Their doors are 50 ft. apart. Each starts at his or her own door and walks at a steady pace toward each other and stops when they meet.

What would their graphing stories look like if we put them on the *same* graph? When the two people meet in the hallway, what would be happening on the graph? Sketch a graph that shows their distance from Maya's door.

Exploratory Challenge/Exercises 1–4

Watch the following graphing story.

http://youtu.be/X956EvmCevI

The video shows a man and a girl walking on the same stairway.

1. Graph the man's elevation on the stairway versus time in seconds.

2. Add the girl's elevation to the same graph. How did you account for the fact that the two people did not start at the same time?

3. Suppose the two graphs intersect at the point $P(24, 4)$. What is the meaning of this point in this situation?

4. Is it possible for two people, walking in stairwells, to produce the same graphs you have been using and not pass each other at time 12 sec.? Explain your reasoning.

Example 2/Exercises 5–7

Consider the story:

Duke starts at the base of a ramp and walks up it at a constant rate. His elevation increases by 3 ft. every second. Just as Duke starts walking up the ramp, Shirley starts at the top of the same 25 ft. high ramp and begins walking down the ramp at a constant rate. Her elevation decreases 2 ft. every second.

5. Sketch two graphs on the same set of elevation-versus-time axes to represent Duke's and Shirley's motions.

6. What are the coordinates of the point of intersection of the two graphs? At what time do Duke and Shirley pass each other?

7. Write down the equation of the line that represents Duke's motion as he moves up the ramp and the equation of the line that represents Shirley's motion as she moves down the ramp. Show that the coordinates of the point you found in the question above satisfy both equations.

EUREKA
MATH™

Lesson Summary

The *intersection point* of the graphs of two equations is an ordered pair that is a solution to both equations. In the context of a distance (or elevation) story, this point represents the fact that both distances (or elevations) are equal at the given time.

Graphing stories with quantities that change at a constant rate can be represented using piecewise linear equations.

Problem Set

1. Reread the story about Maya and Earl from Example 1. Suppose that Maya walks at a constant rate of 3 ft. every second and Earl walks at a constant rate of 4 ft. every second starting from 50 ft. away. Create equations for each person's distance from Maya's door and determine exactly when they meet in the hallway. How far are they from Maya's door at this time?

2. Consider the story:

   May, June, and July were running at the track. May started first and ran at a steady pace of 1 *mi. every* 11 *min. June started* 5 *min. later than May and ran at a steady pace of* 1 *mi. every* 9 *min. July started* 2 *min. after June and ran at a steady pace, running the first lap* $\left(\frac{1}{4}\text{ mi.}\right)$ *in* 1.5 *min. She maintained this steady pace for* 3 *more laps and then slowed down to* 1 *lap every* 3 *min.*

   a. Sketch May, June, and July's distance-versus-time graphs on a coordinate plane.

   b. Create linear equations that represent each girl's mileage in terms of time in minutes. You will need two equations for July since her pace changes after 4 laps (1 mi.).

   c. Who was the first person to run 3 mi.?

   d. Did June and July pass May on the track? If they did, when and at what mileage?

   e. Did July pass June on the track? If she did, when and at what mileage?

3. Suppose two cars are travelling north along a road. Car 1 travels at a constant speed of 50 mph for two hours, then speeds up and drives at a constant speed of 100 mph for the next hour. The car breaks down and the driver has to stop and work on it for two hours. When he gets it running again, he continues driving recklessly at a constant speed of 100 mph. Car 2 starts at the same time that Car 1 starts, but Car 2 starts 100 mi. farther north than Car 1 and travels at a constant speed of 25 mph throughout the trip.

   a. Sketch the distance-versus-time graphs for Car 1 and Car 2 on a coordinate plane. Start with time 0 and measure time in hours.

   b. Approximately when do the cars pass each other?

   c. Tell the entire story of the graph from the point of view of Car 2. (What does the driver of Car 2 see along the way and when?)

   d. Create linear equations representing each car's distance in terms of time (in hours). Note that you will need four equations for Car 1 and only one for Car 2. Use these equations to find the exact coordinates of when the cars meet.

4. Suppose that in Problem 3 above, Car 1 travels at the constant speed of 25 mph the entire time. Sketch the distance-versus-time graphs for the two cars on a graph below. Do the cars ever pass each other? What is the linear equation for Car 1 in this case?

5. Generate six distinct random whole numbers between 2 and 9 inclusive, and fill in the blanks below with the numbers in the order in which they were generated.

$$A \ (0 \ , \_\_\_\_), \quad B \ (\_\_\_\_, \_\_\_\_), \quad C \ (10 \ , \_\_\_\_)$$

$$D \ (0 \ , \_\_\_\_), \quad E \ (10 \ , \_\_\_\_).$$

   (Link to a random number generator http://www.mathgoodies.com/calculators/random_no_custom.html)

   a. On a coordinate plane, plot points A, B, and C. Draw line segments from point A to point B, and from point B to point C.

   b. On the same coordinate plane, plot points D and E and draw a line segment from point D to point E.

   c. Write a graphing story that describes what is happening in this graph. Include a title, x- and y-axis labels, and scales on your graph that correspond to your story.

Lesson 5:   Two Graphing Stories

EUREKA MATH™

6. The following graph shows the revenue (or income) a company makes from designer coffee mugs and the total cost (including overhead, maintenance of machines, etc.) that the company spends to make the coffee mugs.

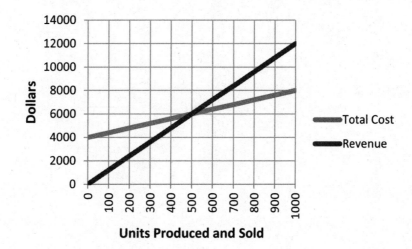

a. How are revenue and total cost related to the number of units of coffee mugs produced?

b. What is the meaning of the point $(0, 4000)$ on the total cost line?

c. What are the coordinates of the intersection point? What is the meaning of this point in this situation?

d. Create linear equations for revenue and total cost in terms of units produced and sold. Verify the coordinates of the intersection point.

e. Profit for selling 1,000 units is equal to revenue generated by selling 1,000 units minus the total cost of making 1,000 units. What is the company's profit if 1,000 units are produced and sold?

This page intentionally left blank

Lesson 6: Algebraic Expressions—The Distributive Property

Classwork

Exercises

2. Using the numbers 1, 2, 3, 4 only once and the operations + or × as many times as you like, write an expression that evaluates to 16. Use this expression and any combination of those symbols as many times as you like to write an expression that evaluates to 816.

3. Define the rules of a game as follows:

    a. Begin by choosing an initial set of symbols, variable or numeric, as a starting set of expressions.

    b. Generate more expressions by placing any previously created expressions into the blanks of the addition operator: _____ + _____.

4. Roma says that collecting like terms can be seen as an application of the distributive property. Is writing $x + x = 2x$ an application of the distributive property?

5. Leela is convinced that $(a + b)^2 = a^2 + b^2$. Do you think she is right? Use a picture to illustrate your reasoning.

6. Draw a picture to represent the expression $(a + b + 1) \times (b + 1)$.

7. Draw a picture to represent the expression $(a + b) \times (c + d) \times (e + f + g)$.

A Key Belief of Arithmetic

THE DISTRIBUTIVE PROPERTY: If a, b, and c are real numbers, then $a(b + c) = ab + ac$.

EUREKA
MATH™

Lesson Summary

The distributive property represents a key belief about the arithmetic of real numbers. This property can be applied to algebraic expressions using variables that represent real numbers.

Problem Set

1. Insert parentheses to make each statement true.
   a. $2 + 3 \times 4^2 + 1 = 81$
   b. $2 + 3 \times 4^2 + 1 = 85$
   c. $2 + 3 \times 4^2 + 1 = 51$
   d. $2 + 3 \times 4^2 + 1 = 53$

2. Using starting symbols of w, q, 2, and -2, which of the following expressions will NOT appear when following the rules of the game played in Exercise 3?
   a. $7w + 3q + (-2)$
   b. $q - 2$
   c. $w - q$
   d. $2w + 6$
   e. $-2w + 2$

3. Luke wants to play the 4-number game with the numbers 1, 2, 3, and 4 and the operations of addition, multiplication, AND subtraction.

   Leoni responds, "Or we just could play the 4-number game with just the operations of addition and multiplication, but now with the numbers -1, -2, -3, -4, 1, 2, 3, and 4 instead."

   What observation is Leoni trying to point out to Luke?

4. Consider the expression: $(x + 3) \cdot (y + 1) \cdot (x + 2)$.
   a. Draw a picture to represent the expression.
   b. Write an equivalent expression by applying the distributive property.

5.
   a. Given that $a > b$, which of the shaded regions is larger and why?

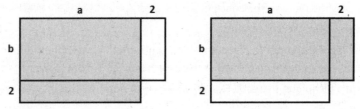

b. Consider the expressions 851×29 and 849×31. Which would result in a larger product? Use a diagram to demonstrate your result.

6. Consider the following diagram.

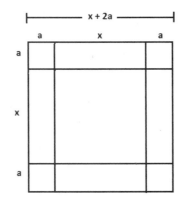

Edna looked at the diagram and then highlighted the four small rectangles shown and concluded: $(x + 2a)^2 = x^2 + 4a(x + a)$.

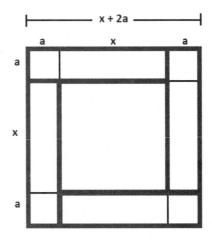

a. Michael, when he saw the picture, highlighted four rectangles and concluded:

$(x + 2a)^2 = x^2 + 2ax + 2a(x + 2a)$.

Which four rectangles and one square did he highlight?

b. Jill, when she saw the picture, highlighted eight rectangles and squares (not including the square in the middle) to conclude:

$(x + 2a)^2 = x^2 + 4ax + 4a^2$.

Which eight rectangles and squares did she highlight?

c. When Fatima saw the picture, she concluded:

$(x + 2a)^2 = x^2 + 4a(x + 2a) - 4a^2$.

She claims she highlighted just four rectangles to conclude this. Identify the four rectangles she highlighted, and explain how using them she arrived at the expression $x^2 + 4a(x + 2a) - 4a^2$.

d. Is each student's technique correct? Explain why or why not.

Lesson 6:   Algebraic Expressions—The Distributive Property

EUREKA MATH™

Lesson 7: Algebraic Expressions—The Commutative and Associative Properties

Classwork

Exercise 1

Suzy draws the following picture to represent the sum $3 + 4$:

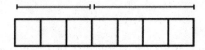

Ben looks at this picture from the opposite side of the table and says, "You drew $4 + 3$."

Explain why Ben might interpret the picture this way.

$4 + 3$

Exercise 2

Suzy adds more to her picture and says, "The picture now represents $(3 + 4) + 2$."

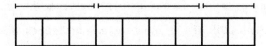

How might Ben interpret this picture? Explain your reasoning.

Exercise 3

Suzy then draws another picture of squares to represent the product 3×4. Ben moves to the end of the table and says, "From my new seat, your picture looks like the product 4×3."

What picture might Suzy have drawn? Why would Ben see it differently from his viewpoint?

Exercise 4

Draw a picture to represent the quantity $(3 \times 4) \times 5$ that also could represent the quantity $(4 \times 5) \times 3$ when seen from a different viewpoint.

Four Properties of Arithmetic:

THE COMMUTATIVE PROPERTY OF ADDITION: If a and b are real numbers, then $a + b = b + a$.

THE ASSOCIATIVE PROPERTY OF ADDITION: If a, b, and c are real numbers, then $(a + b) + c = a + (b + c)$.

THE COMMUTATIVE PROPERTY OF MULTIPLICATION: If a and b are real numbers, then $a \times b = b \times a$.

THE ASSOCIATIVE PROPERTY OF MULTIPLICATION: If a, b, and c are real numbers, then $(ab)c = a(bc)$.

Exercise 5

Viewing the diagram below from two different perspectives illustrates that $(3 + 4) + 2$ equals $2 + (4 + 3)$.

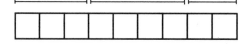

Is it true for all real numbers x, y, and z that $(x + y) + z$ should equal $(z + y) + x$?

(Note: The direct application of the associative property of addition only gives $(x + y) + z = x + (y + z)$.)

EUREKA
MATH™

Exercise 6

Draw a flow diagram and use it to prove that $(xy)z = (zy)x$ for all real numbers x, y, and z.

Exercise 7

Use these abbreviations for the properties of real numbers, and complete the flow diagram.

C_+ for the commutative property of addition

$C_\times$ for the commutative property of multiplication

A_+ for the associative property of addition

$A_\times$ for the associative property of multiplication

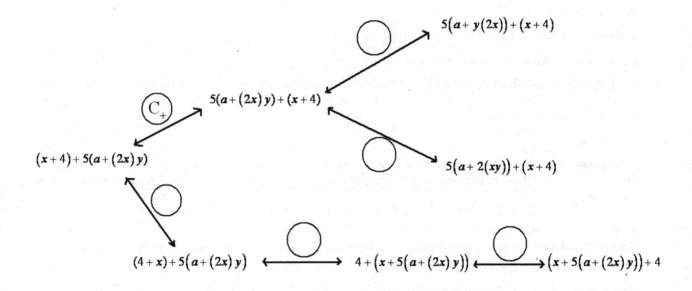

Exercise 8

Let a, b, c, and d be real numbers. Fill in the missing term of the following diagram to show that $((a + b) + c) + d$ is sure to equal $a + (b + (c + d))$.

$$((a+b)+c)+d \xleftrightarrow{\ A\ } (a+(b+c))+d \xleftrightarrow{\ A\ } \boxed{} \xleftrightarrow{\ A\ } a+(b+(c+d))$$

NUMERICAL SYMBOL: A *numerical symbol* is a symbol that represents a specific number.

For example, $0, 1, 2, 3, \frac{2}{3}, -3, -124.122, \pi, e$ are numerical symbols used to represent specific points on the real number line.

VARIABLE SYMBOL: A *variable symbol* is a symbol that is a placeholder for a number.

It is possible that a question may restrict the type of number that a placeholder might permit (e.g., integers only or positive real numbers).

ALGEBRAIC EXPRESSION: An *algebraic expression* is either

1. A numerical symbol or a variable symbol, or
2. The result of placing previously generated algebraic expressions into the two blanks of one of the four operators $((\_) + (\_)$, $(\_) - (\_)$, $(\_) \times (\_)$, $(\_) \div (\_))$ or into the base blank of an exponentiation with an exponent that is a rational number.

Two algebraic expressions are *equivalent* if we can convert one expression into the other by repeatedly applying the commutative, associative, and distributive properties and the properties of rational exponents to components of the first expression.

NUMERICAL EXPRESSION: A *numerical expression* is an algebraic expression that contains only numerical symbols (no variable symbols), which evaluate to a single number.

The expression $3 \div 0$, is not a numerical expression.

EQUIVALENT NUMERICAL EXPRESSIONS: Two numerical expressions are *equivalent* if they evaluate to the same number.

Note that $1 + 2 + 3$ and $1 \times 2 \times 3$, for example, are equivalent numerical expressions (they are both 6), but $a + b + c$ and $a \times b \times c$ are not equivalent expressions.

> **Lesson Summary**
>
> The commutative and associative properties represent key beliefs about the arithmetic of real numbers. These properties can be applied to algebraic expressions using variables that represent real numbers.
>
> Two algebraic expressions are **equivalent** if we can convert one expression into the other by repeatedly applying the commutative, associative, and distributive properties and the properties of rational exponents to components of the first expression.

Problem Set

1. The following portion of a flow diagram shows that the expression $ab + cd$ is equivalent to the expression $dc + ba$.

Fill in each circle with the appropriate symbol: Either C_+ (for the commutative property of addition) or $C_\times$ (for the commutative property of multiplication).

2. Fill in the blanks of this proof showing that $(w + 5)(w + 2)$ is equivalent to $w^2 + 7w + 10$. Write either commutative property, associative property, or distributive property in each blank.

$$(w + 5)(w + 2)$$

$= (w + 5)w + (w + 5) \times 2$ _____

$= w(w + 5) + (w + 5) \times 2$ _____

$= w(w + 5) + 2(w + 5)$ _____

$= w^2 + w \times 5 + 2(w + 5)$ _____

$= w^2 + 5w + 2(w + 5)$ _____

$= w^2 + 5w + 2w + 10$ _____

$= w^2 + (5w + 2w) + 10$ _____

$= w^2 + 7w + 10$

3. Fill in each circle of the following flow diagram with one of the letters: C for commutative property (for either addition or multiplication), A for associative property (for either addition or multiplication), or D for distributive property.

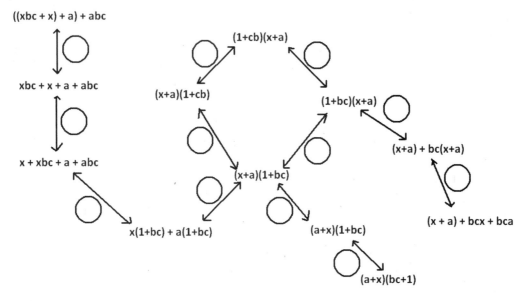

4. What is a quick way to see that the value of the sum $53 + 18 + 47 + 82$ is 200?

5.

    a. If $ab = 37$ and $= \frac{1}{37}$, what is the value of the product $x \times b \times y \times a$?

    b. Give some indication as to how you used the commutative and associative properties of multiplication to evaluate $x \times b \times y \times a$ in part (a).

    c. Did you use the associative and commutative properties of addition to answer Question 4?

6. The following is a proof of the algebraic equivalency of $(2x)^3$ and $8x^3$. Fill in each of the blanks with either the statement *commutative property* or *associative property*.

$$
\begin{aligned}
(2x)^3 &= 2x \cdot 2x \cdot 2x \\
&= 2(x \times 2)(x \times 2)x \qquad \underline{\hspace{6cm}} \\
&= 2(2x)(2x)x \qquad \underline{\hspace{6cm}} \\
&= 2 \cdot 2(x \times 2)x \cdot x \qquad \underline{\hspace{6cm}} \\
&= 2 \cdot 2(2x)x \cdot x \qquad \underline{\hspace{6cm}} \\
&= (2 \cdot 2 \cdot 2)(x \cdot x \cdot x) \qquad \underline{\hspace{6cm}} \\
&= 8x^3
\end{aligned}
$$

7. Write a mathematical proof of the algebraic equivalency of $(ab)^2$ and a^2b^2.

8.

   a. Suppose we are to play the 4-number game with the symbols a, b, c, and d to represent numbers, each used at most once, combined by the operation of addition ONLY. If we acknowledge that parentheses are unneeded, show there are essentially only 15 expressions one can write.

   b. How many answers are there for the multiplication ONLY version of this game?

9. Write a mathematical proof to show that $(x + a)(x + b)$ is equivalent to $x^2 + ax + bx + ab$.

10. Recall the following rules of exponents:

$$x^a \cdot x^b = x^{a+b} \qquad\qquad \frac{x^a}{x^b} = x^{a-b} \qquad\qquad (x^a)^b = x^{ab}$$

$$(xy)^a = x^a y^a \qquad\qquad \left(\frac{x}{y}\right)^a = \frac{x^a}{y^a}$$

Here x, y, a, and b are real numbers with x and y nonzero.

Replace each of the following expressions with an equivalent expression in which the variable of the expression appears only once with a positive number for its exponent. (For example, $\frac{7}{b^2} \cdot b^{-4}$ is equivalent to $\frac{7}{b^6}$.)

   a. $(16x^2) \div (16x^5)$
   b. $(2x)^4(2x)^3$
   c. $(9z^{-2})(3z^{-1})^{-3}$
   d. $\left((25w^4) \div (5w^3)\right) \div (5w^{-7})$
   e. $(25w^4) \div \left((5w^3) \div (5w^{-7})\right)$

Optional Challenge:

11. Grizelda has invented a new operation that she calls the *average operator*. For any two real numbers a and b, she declares $a \oplus b$ to be the average of a and b:

$$a \oplus b = \frac{a + b}{2}$$

   a. Does the average operator satisfy a commutative property? That is, does $a \oplus b = b \oplus a$ for all real numbers a and b?

   b. Does the average operator distribute over addition? That is, does $a \oplus (b + c) = (a \oplus b) + (a \oplus c)$ for all real numbers a, b, and c?

This page intentionally left blank

Lesson 8: Adding and Subtracting Polynomials

Classwork

Exercise 1

a. How many quarters, nickels, and pennies are needed to make $1.13?

b. Fill in the blanks:

$8{,}943 = \underline{8} \times 1000 + \underline{9} \times 100 + \underline{4} \times 10 + \underline{3} \times 1$

$= \underline{8} \times 10^3 + \underline{9} \times 10^2 + \underline{4} \times 10 + \underline{3} \times 1$

c. Fill in the blanks:

$8{,}943 = \underline{\hspace{2em}} \times 20^3 + \underline{\hspace{2em}} \times 20^2 + \underline{\hspace{2em}} \times 20 + \underline{\hspace{2em}} \times 1$

d. Fill in the blanks:

$113 = \underline{\hspace{2em}} \times 5^2 + \underline{\hspace{2em}} \times 5 + \underline{\hspace{2em}} \times 1$

Exercise 2

Now let's be as general as possible by not identifying which base we are in. Just call the base x.

Consider the expression $1 \cdot x^3 + 2 \cdot x^2 + 7 \cdot x + 3 \cdot 1$, or equivalently $x^3 + 2x^2 + 7x + 3$.

a. What is the value of this expression if $x = 10$?

b. What is the value of this expression if $x = 20$?

Exercise 3

a. When writing numbers in base 10, we only allow coefficients of 0 through 9. Why is that?

b. What is the value of $22x + 3$ when $x = 5$? How much money is 22 nickels and 3 pennies?

c. What number is represented by $4x^2 + 17x + 2$ if $x = 10$?

d. What number is represented by $4x^2 + 17x + 2$ if $x = -2$ or if $x = \frac{2}{3}$?

e. What number is represented by $-3x^2 + \sqrt{2}x + \frac{1}{2}$ when $x = \sqrt{2}$?

POLYNOMIAL EXPRESSION: A *polynomial expression* is either

1. A numerical expression or a variable symbol, or

2. The result of placing two previously generated polynomial expressions into the blanks of the addition operator (__+__) or the multiplication operator (__×__).

EUREKA
MATH™

Exercise 4

Find each sum or difference by combining the parts that are alike.

a. $417 + 231 =$ _____ hundreds + _____ tens + _____ ones + _____ hundreds + _____ tens + _____ ones

$\quad\quad\quad\quad\quad = $ _____ hundreds + _____ tens + _____ ones

b. $(4x^2 + x + 7) + (2x^2 + 3x + 1)$

c. $(3x^3 - x^2 + 8) - (x^3 + 5x^2 + 4x - 7)$

d. $3(x^3 + 8x) - 2(x^3 + 12)$

e. $(5 - t - t^2) + (9t + t^2)$

f. $(3p + 1) + 6(p - 8) - (p + 2)$

Lesson Summary

A *monomial* is a polynomial expression generated using only the multiplication operator ($\_\times\_$). Thus, it does not contain + or − operators. *Monomials* are written with numerical factors multiplied together and variable or other symbols each occurring one time (using exponents to condense multiple instances of the same variable).

A *polynomial* is the sum (or difference) of monomials.

The *degree of a monomial* is the sum of the exponents of the variable symbols that appear in the monomial.

The *degree of a polynomial* is the degree of the monomial term with the highest degree.

Problem Set

1. Celina says that each of the following expressions is actually a binomial in disguise:

   i. $5abc - 2a^2 + 6abc$

   ii. $5x^3 \cdot 2x^2 - 10x^4 + 3x^5 + 3x \cdot (-2)x^4$

   iii. $(t + 2)^2 - 4t$

   iv. $5(a - 1) - 10(a - 1) + 100(a - 1)$

   v. $(2\pi r - \pi r^2)r - (2\pi r - \pi r^2) \cdot 2r$

   For example, she sees that the expression in (i) is algebraically equivalent to $11abc - 2a^2$, which is indeed a binomial. (She is happy to write this as $11abc + (-2)a^2$, if you prefer.)

   Is she right about the remaining four expressions?

2. Janie writes a polynomial expression using only one variable, x, with degree 3. Max writes a polynomial expression using only one variable, x, with degree 7.

   a. What can you determine about the degree of the sum of Janie's and Max's polynomials?

   b. What can you determine about the degree of the difference of Janie's and Max's polynomials?

3. Suppose Janie writes a polynomial expression using only one variable, x, with degree of 5, and Max writes a polynomial expression using only one variable, x, with degree of 5.

   a. What can you determine about the degree of the sum of Janie's and Max's polynomials?

   b. What can you determine about the degree of the difference of Janie's and Max's polynomials?

4. Find each sum or difference by combining the parts that are alike.

   a. $(2p + 4) + 5(p - 1) - (p + 7)$

   b. $(7x^4 + 9x) - 2(x^4 + 13)$

   c. $(6 - t - t^4) + (9t + t^4)$

   d. $(5 - t^2) + 6(t^2 - 8) - (t^2 + 12)$

   e. $(8x^3 + 5x) - 3(x^3 + 2)$

   f. $(12x + 1) + 2(x - 4) - (x - 15)$

   g. $(13x^2 + 5x) - 2(x^2 + 1)$

   h. $(9 - t - t^2) - \frac{3}{2}(8t + 2t^2)$

   i. $(4m + 6) - 12(m - 3) + (m + 2)$

   j. $(15x^4 + 10x) - 12(x^4 + 4x)$

Lesson 8:   Adding and Subtracting Polynomials

EUREKA MATH™

Lesson 9: Multiplying Polynomials

Classwork

Exercise 1

a.   Gisella computed 342×23 as follows:

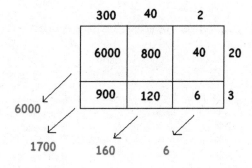

Can you explain what she is doing? What is her final answer?

Use a geometric diagram to compute the following products:

b. $(3x^2 + 4x + 2)(2x + 3)$

c. $(2x^2 + 10x + 1)(x^2 + x + 1)$

d. $(x-1)(x^3 + 6x^2 - 5)$

Exercise 2

Multiply the polynomials using the distributive property: $(3x^2 + x - 1)(x^4 - 2x + 1)$.

Exercise 3

The expression $10x^2 + 6x^3$ is the result of applying the distributive property to the expression $2x^2(5 + 3x)$. It is also the result of applying the distributive property to $2(5x^2 + 3x^3)$ or to $x(10x + 6x^2)$, for example, or even to $1 \cdot (10x^2 + 6x^3)$.

For (a) to (j) below, write down an expression such that if you applied the distributive property to your expression, it would give the result presented. Give interesting answers!

   a. $6a + 14a^2$

   b. $2x^4 + 2x^5 + 2x^{10}$

Multiplying Polynomials

EUREKA MATH™

c. $6z^2 - 15z$

d. $42w^3 - 14w + 77w^5$

e. $z^2(a + b) + z^3(a + b)$

f. $\dfrac{3}{2}s^2 + \dfrac{1}{2}$

g. $15p^3r^4 - 6p^2r^5 + 9p^4r^2 + 3\sqrt{2}p^3r^6$

h. $0.4x^9 - 40x^8$

i. $(4x + 3)(x^2 + x^3) - (2x + 2)(x^2 + x^3)$

j. $(2z + 5)(z - 2) - (13z - 26)(z - 3)$

Exercise 4

Sammy wrote a polynomial using only one variable, x, of degree 3. Myisha wrote a polynomial in the same variable of degree 5. What can you say about the degree of the product of Sammy's and Myisha's polynomials?

Extension

Find a polynomial that, when multiplied by $2x^2 + 3x + 1$, gives the answer $2x^3 + x^2 - 2x - 1$.

Problem Set

1. Use the distributive property to write each of the following expressions as the sum of monomials.

   a. $3a(4 + a)$

   b. $x(x + 2) + 1$

   c. $\frac{1}{3}(12z + 18z^2)$

   d. $4x(x^3 - 10)$

   e. $(x - 4)(x + 5)$

   f. $(2z - 1)(3z^2 + 1)$

   g. $(10w - 1)(10w + 1)$

   h. $(-5w - 3)w^2$

   i. $16s^{100}\left(\frac{1}{2}s^{200} + 0.125s\right)$

   j. $(2q + 1)(2q^2 + 1)$

   k. $(x^2 - x + 1)(x - 1)$

   l. $3xz(9xy + z) - 2yz(x + y - z)$

   m. $(t - 1)(t + 1)(t^2 + 1)$

   n. $(w + 1)(w^4 - w^3 + w^2 - w + 1)$

   o. $z(2z + 1)(3z - 2)$

   p. $(x + y)(y + z)(z + x)$

   q. $\frac{x+y}{3}$

   r. $(20f^{10} - 10f^5) \div 5$

   s. $-5y(y^2 + y - 2) - 2(2 - y^3)$

   t. $\frac{(a+b-c)(a+b+c)}{17}$ sd

   u. $(2x \div 9 + (5x) \div 2) \div (-2)$

   v. $(-2f^3 - 2f + 1)(f^2 - f + 2)$

2. Use the distributive property (and your wits!) to write each of the following expressions as a sum of monomials. If the resulting polynomial is in one variable, write the polynomial in standard form.

   a. $(a + b)^2$

   b. $(a + 1)^2$

   c. $(3 + b)^2$

   d. $(3 + 1)^2$

   e. $(x + y + z)^2$

   f. $(x + 1 + z)^2$

   g. $(3 + z)^2$

   h. $(p + q)^3$

   i. $(p - 1)^3$

   j. $(5 + q)^3$

3. Use the distributive property (and your wits!) to write each of the following expressions as a polynomial in standard form.

   a. $(s^2 + 4)(s - 1)$

   b. $3(s^2 + 4)(s - 1)$

   c. $s(s^2 + 4)(s - 1)$

   d. $(s + 1)(s^2 + 4)(s - 1)$

   e. $(u - 1)(u^5 + u^4 + u^3 + u^2 + u + 1)$

   f. $\sqrt{5}(u - 1)(u^5 + u^4 + u^3 + u^2 + u + 1)$

   g. $(u^7 + u^3 + 1)(u - 1)(u^5 + u^4 + u^3 + u^2 + u + 1)$

4. Beatrice writes down every expression that appears in this problem set, one after the other, linking them with + signs between them. She is left with one very large expression on her page. Is that expression a polynomial expression? That is, is it algebraically equivalent to a polynomial?

What if she wrote − signs between the expressions instead?

What if she wrote × signs between the expressions instead?

Lesson 10: True and False Equations

Classwork

Exercise 1

a.   Consider the statement: "The president of the United States is a United States citizen."
     Is the statement a grammatically correct sentence?
     What is the subject of the sentence? What is the verb in the sentence?
     Is the sentence true?

b.   Consider the statement: "The president of France is a United States citizen."
     Is the statement a grammatically correct sentence?
     What is the subject of the sentence? What is the verb in the sentence?
     Is the sentence true?

c.   Consider the statement: "$2 + 3 = 1 + 4$."
     This is a sentence. What is the verb of the sentence? What is the subject of the sentence?
     Is the sentence true?

d.   Consider the statement: "$2 + 3 = 9 + 4$."
     Is this statement a sentence? And if so, is the sentence true or false?

A *number sentence* is a statement of equality between two numerical expressions.

A *number sentence* is said to be *true* if both numerical expressions are equivalent (that is, both evaluate to the same number). It is said to be *false* otherwise. True and false are called *truth values*.

Lesson 10:   True and False Equations

S.49

Exercise 2

Determine whether the following number sentences are true or false.

a. $4 + 8 = 10 + 5$

b. $\dfrac{1}{2} + \dfrac{5}{8} = 1.2 - 0.075$

c. $(71 \cdot 603) \cdot 5876 = 603 \cdot (5876 \cdot 71)$

d. $13 \times 175 = 13 \times 90 + 85 \times 13$

e. $(7 + 9)^2 = 7^2 + 9^2$

f. $\pi = 3.141$

g. $\sqrt{(4 + 9)} = \sqrt{4} + \sqrt{9}$

EUREKA
MATH™

h. $\dfrac{1}{2} + \dfrac{1}{3} = \dfrac{2}{5}$

i. $\dfrac{1}{2} + \dfrac{1}{3} = \dfrac{2}{6}$

j. $\dfrac{1}{2} + \dfrac{1}{3} = \dfrac{5}{6}$

k. $3^2 + 4^2 = 7^2$

l. $3^2 \times 4^2 = 12^2$

m. $3^2 \times 4^3 = 12^6$

n. $3^2 \times 3^3 = 3^5$

Exercise 3

    a.   Could a number sentence be both true and false?

    b.   Could a number sentence be neither true nor false?

An *algebraic equation* is a statement of equality between two expressions.

Algebraic equations can be number sentences (when both expressions are numerical), but often they contain symbols whose values have not been determined.

Exercise 4

    a.   Which of the following are algebraic equations?

        i. $3.1x - 11.2 = 2.5x + 2.3$

        ii. $10\pi^4 + 3 = 99\pi^2$

        iii. $\pi + \pi = 2\pi$

        iv. $\dfrac{1}{2} + \dfrac{1}{2} = \dfrac{2}{4}$

        v. $79\pi^3 + 70\pi^2 - 56\pi + 87 = \dfrac{60\pi + 29\,928}{\pi^2}$

    b.   Which of them are also number sentences?

        Lesson 10:   True and False Equations

EUREKA MATH

c.   For each number sentence, state whether the number sentence is true or false.

Exercises 5

When algebraic equations contain a symbol whose value has not yet been determined, we use analysis to determine whether:

a.   The equation is true for all the possible values of the variable(s), or

b.   The equation is true for a certain set of the possible value(s) of the variable(s), or

c.   The equation is never true for any of the possible values of the variable(s).

For each of the three cases, write an algebraic equation that would be correctly described by that case. Use only the variable, x, where x represents a real number.

Example 1

Consider the following scenario.

Julie is 300 feet away from her friend's front porch and observes, "Someone is sitting on the porch."

Given that she did not specify otherwise, we would assume that the *someone* Julie thinks she sees is a human. We cannot guarantee that Julie's observational statement is true. It could be that Julie's friend has something on the porch that merely looks like a human from far away. Julie assumes she is correct and moves closer to see if she can figure out who it is. As she nears the porch, she declares, "Ah, it is our friend, John Berry."

Exercise 6

Name a value of the variable that would make each equation a true number sentence.

Here are several examples of how we can name the value of the variable:

<u>Let $w = -2$</u>. Then, $w^2 = 4$ is true.

$w^2 = 4$ is true <u>when $w = -2$.</u>

$w^2 = 4$ is true <u>if $w = -2$.</u>

$w^2 = 4$ is true <u>for $w = -2$ and $w = 2$.</u>

There might be more than one option for what numerical values to write. (And feel free to write more than one possibility.)

Warning: Some of these are tricky. Keep your wits about you!

a. Let _____. Then, $7 + x = 12$ is true.

b. Let _____. Then, $3r + 0.5 = \frac{37}{2}$ is true.

c. $m^3 = -125$ is true for _____.

d. A number x and its square, x^2, have the same value when _____.

e. The average of 7 and n is -8 if _____.

f. Let _____. Then, $2a = a + a$ is true.

g. $q + 67 = q + 68$ is true for _____.

Lesson 10:   True and False Equations

EUREKA
MATH™

Problem Set

Determine whether the following number sentences are true or false.

1. $18 + 7 = \dfrac{50}{2}$

2. $3.123 = 9.369 \cdot \dfrac{1}{3}$

3. $(123 + 54) \cdot 4 = 123 + (54 \cdot 4)$

4. $5^2 + 12^2 = 13^2$

5. $(2 \times 2)^2 = \sqrt{256}$

6. $\dfrac{4}{3} = 1.333$

In the following equations, let $x = -3$ and $y = \dfrac{2}{3}$. Determine whether the following equations are true, false, or neither true nor false.

7. $xy = -2$

8. $x + 3y = -1$

9. $x + z = 4$

10. $9y = -2x$

11. $\dfrac{y}{x} = -2$

12. $\dfrac{-\frac{2}{x}}{y} = -1$

For each of the following, assign a value to the variable, x, to make the equation a true statement.

13. $(x^2 + 5)(3 + x^4)(100x^2 - 10)(100x^2 + 10) = 0$ for _____.

14. $\sqrt{(x+1)(x+2)} = \sqrt{20}$ for _____.

15. $(d + 5)^2 = 36$ for _____.

16. $(2z + 2)(z^5 - 3) + 6 = 0$ for _____.

17. $\dfrac{1+x}{1+x^2} = \dfrac{3}{5}$ for _____.

18. $\dfrac{1+x}{1+x^2} = \dfrac{2}{5}$ for _____.

19. The diagonal of a square of side length L is 2 inches long when _____.

20. $\left(T - \sqrt{3}\right)^2 = T^2 + 3$ for _____.

21. $\dfrac{1}{x} = \dfrac{x}{1}$ if _____.

22. $\left(2 + \left(2 - \left(2 + \left(2 - (2 + r)\right)\right)\right)\right) = 1$ for _____.

23. $x + 2 = 9$

24. $x + 2^2 = -9$

25. $-12t = 12$

26. $12t = 24$

27. $\dfrac{1}{b-2} = \dfrac{1}{4}$

28. $\dfrac{1}{2b-2} = -\dfrac{1}{4}$

29. $\sqrt{x} + \sqrt{5} = \sqrt{x+5}$

30. $(x-3)^2 = x^2 + (-3)^2$

31. $x^2 = -49$

32. $\dfrac{2}{3} + \dfrac{1}{5} = \dfrac{3}{x}$

Fill in the blank with a variable term so that the given value of the variable will make the equation true.

33. _____ $+ 4 = 12; x = 8$

34. _____ $+ 4 = 12; x = 4$

Fill in the blank with a constant term so that the given value of the variable will make the equation true.

35. $4y -$ ____ $= 100; y = 25$

36. $4y -$ ____ $= 0; y = 6$

37. $r +$ ____ $= r; r$ is any real number.

38. $r \times$ ____ $= r; r$ is any real number.

Generate the following:

39. An equation that is always true

40. An equation that is true when $x = 0$

41. An equation that is never true

42. An equation that is true when $t = 1$ or $t = -1$

43. An equation that is true when $y = -0.5$

44. An equation that is true when $z = \pi$

EUREKA
MATH

Lesson 11: Solution Sets for Equations and Inequalities

Classwork

Example 1

Consider the equation, $x^2 = 3x + 4$, where x represents a real number.

a.   Are the expressions x^2 and $3x + 4$ algebraically equivalent?

b.   The following table shows how we might "sift" through various values to assign to the variable symbol x in the hunt for values that would make the equation true.

x-VALUE	THE EQUATION	TRUTH VALUE
Let $x = 0$	$0^2 = 3(0) + 4$	FALSE
Let $x = 5$	$5^2 = 3(5) + 4$	FALSE
Let $x = 6$	$6^2 = 3(6) + 4$	FALSE
Let $x = -7$	$(-7)^2 = 3(-7) + 4$	FALSE
Let $x = 4$	$4^2 = 3(4) + 4$	TRUE
Let $x = 9$	$9^2 = 3(9) + 4$	FALSE
Let $x = 10$	$10^2 = 3(10) + 4$	FALSE
Let $x = -8$	$(-8)^2 = 3(-8) + 4$	FALSE

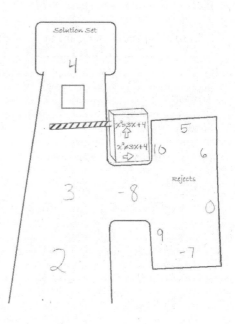

Example 2

Consider the equation $7 + p = 12$.

p-VALUE	THE NUMBER SENTENCE	TRUTH VALUE
Let $p = 0$	$7 + 0 = 12$	FALSE
Let $p = 4$		
Let $p = 1 + \sqrt{2}$		
Let $p = \dfrac{1}{\pi}$		
Let $p = 5$		

The *solution set* of an equation written with only one variable is the set of all values one can assign to that variable to make the equation a true statement. Any one of those values is said to be a *solution to the equation*.

To *solve an equation* means to *find the solution set* for that equation.

Example 3

Solve for a: $a^2 = 25$.

One can describe a solution set in any of the following ways:

IN WORDS: $a^2 = 25$ has solutions 5 and -5. ($a^2 = 25$ is true when $a = 5$ or $a = -5$.)

IN SET NOTATION: The solution set of $a^2 = 25$ is $\{-5, 5\}$.

IN A GRAPHICAL REPRESENTATION ON A NUMBER LINE: The solution set of $a^2 = 25$ is

In this graphical representation, a solid dot is used to indicate a point on the number line that is to be included in the solution set. (WARNING: The dot one physically draws is larger than the point it represents. One hopes that it is clear from the context of the diagram which point each dot refers to.)

How set notation works.

- The curly brackets { } indicate we are denoting a set. A set is essentially a collection of things (e.g., letters, numbers, cars, people). In this case, the things are numbers.

- From this example, the numbers -5 and 5 are called elements of the set. No other elements belong in this particular set because no other numbers make the equation $a^2 = 25$ true.

- When elements are listed, they are listed in increasing order.

- Sometimes, a set is empty; it has no elements. In which case, the set looks like { }. We often denote this with the symbol, ∅. We refer to this as the empty set or the null set.

Exercise 1

Solve for a: $a^2 = -25$. Present the solution set in words, in set notation, and graphically.

Exercise 2

Depict the solution set of $7 + p = 12$ in words, in set notation, and graphically.

Example 4

Solve $\dfrac{x}{x} = 1$ for x over the set of positive real numbers. Depict the solution set in words, in set notation, and graphically.

x-VALUE	THE EQUATION	TRUTH VALUE
Let $x = 2$	$\dfrac{2}{2} = 1$	TRUE
Let $x = 7$	$\dfrac{7}{7} = 1$	TRUE
Let $x = 0.01$	$\dfrac{0.01}{0.01} = 1$	TRUE
Let $x = 562\frac{2}{3}$	$\dfrac{562\frac{2}{3}}{562\frac{2}{3}} = 1$	TRUE
Let $x = 10^{100}$	$\dfrac{10^{100}}{10^{100}} = 1$	TRUE
Let $x = \pi$	$\dfrac{\pi}{\pi} = 1$	TRUE

EUREKA MATH™

Exercise 3

Solve $\dfrac{x}{x} = 1$ for x over the set of all nonzero real numbers. Describe the solution set in words, in set notation, and graphically.

Example 5

Example 5

Solve for x: $x(3 + x) = 3x + x^2$.

Exercise 4

Solve for α: $\alpha + \alpha^2 = \alpha(\alpha + 1)$. Describe carefully the reasoning that justifies your solution. Describe the solution set in words, in set notation, and graphically.

An *identity* is an equation that is always true.

Exercise 5

Identify the properties of arithmetic that justify why each of the following equations has a solution set of all real numbers.

a. $2x^2 + 4x = 2(x^2 + 2x)$

b. $2x^2 + 4x = 4x + 2x^2$

c. $2x^2 + 4x = 2x(2 + x)$

Exercise 6

Create an expression for the right side of each equation such that the solution set for the equation will be all real numbers. (There is more than one possibility for each expression. Feel free to write several answers for each one.)

a. $2x - 5 = $ _____

b. $x^2 + x = $ _____

c. $4 \cdot x \cdot y \cdot z = $ _____

d. $(x + 2)^2 = $ _____

EUREKA
MATH™

Example 6

Solve for w: $w + 2 > 4$.

Exercise 7

a.   Solve for B: $B^2 \geq 9$. Describe the solution set using a number line.

b.   What is the solution set to the statement: "Sticks of lengths 2 yards, 2 yards, and L yards make an isosceles triangle"? Describe the solution set in words and on a number line.

Lesson Summary

The *solution set* of an equation written with only one variable symbol is the set of all values one can assign to that variable to make the equation a true number sentence. Any one of those values is said to be a *solution to the equation*.

To *solve an equation* means to *find the solution set* for that equation.

One can describe a solution set in any of the following ways:

IN WORDS: $a^2 = 25$ has solutions 5 and -5. ($a^2 = 25$ is true when $a = 5$ or $a = -5$.)

IN SET NOTATION: The solution set of $a^2 = 25$ is $\{-5, 5\}$.

It is awkward to express the set of infinitely many numbers in set notation. In these cases we can use the notation {variable symbol number type | a description}. For example $\{x \text{ real} \mid x > 0\}$ reads, "x is a real number where x is greater than zero." The symbol $\mathbb{R}$ can be used to indicate all real numbers.

IN A GRAPHICAL REPRESENTAION ON A NUMBER LINE: The solution set of $a^2 = 25$ is as follows:

In this graphical representation, a solid dot is used to indicate a point on the number line that is to be included in the solution set. (WARNING: The dot one physically draws is larger than the point it represents! One hopes that it is clear from the context of the diagram which point each dot refers to.)

Problem Set

For each solution set graphed below, (a) describe the solution set in words, (b) describe the solution set in set notation, and (c) write an equation or an inequality that has the given solution set.

1.

2.

3.

4.

5.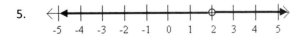

6.

EUREKA MATH

7.    8.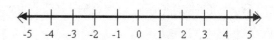

Fill in the chart below.

	SOLUTION SET IN WORDS	SOLUTION SET IN SET NOTATION	GRAPH
9. $z = 2$			
10. $z^2 = 4$			
11. $4z \neq 2$			
12. $z - 3 = 2$			
13. $z^2 + 1 = 2$			
14. $z = 2z$			
15. $z > 2$			
16. $z - 6 = z - 2$			
17. $z - 6 < -2$			
18. $4(z - 1) > 4z - 4$			

For Problems 19–24, answer the following: Are the two expressions algebraically equivalent? If so, state the property (or properties) displayed. If not, state why (the solution set may suffice as a reason) and change the equation, ever so slightly (e.g., touch it up) to create an equation whose solution set is all real numbers.

19. $x(4 - x^2) = (-x^2 + 4)x$

20. $\dfrac{2x}{2x} = 1$

21. $(x - 1)(x + 2) + (x - 1)(x - 5) = (x - 1)(2x - 3)$

22. $\dfrac{x}{5} + \dfrac{x}{3} = \dfrac{2x}{8}$

23. $x^2 + 2x^3 + 3x^4 = 6x^9$

24. $x^3 + 4x^2 + 4x = x(x + 2)^2$

25. Solve for w: $\dfrac{6w+1}{5} \neq 2$. Describe the solution set in set notation.

26. Edwina has two sticks: one 2 yards long and the other 2 meters long. She is going to use them, with a third stick of some positive length, to make a triangle. She has decided to measure the length of the third stick in units of feet.

   a.   What is the solution set to the statement: "Sticks of lengths 2 yards, 2 meters, and L feet make a triangle"? Describe the solution set in words and through a graphical representation.

   b.   What is the solution set to the statement: "Sticks of lengths 2 yards, 2 meters, and L feet make an isosceles triangle"? Describe the solution set in words and through a graphical representation.

   c.   What is the solution set to the statement: "Sticks of lengths 2 yards, 2 meters, and L feet make an equilateral triangle"? Describe the solution set in words and through a graphical representation.

EUREKA MATH

Lesson 12: Solving Equations

Opening Exercise

Answer the following questions.

a. Why should the equations $(x - 1)(x + 3) = 17 + x$ and $(x - 1)(x + 3) = x + 17$ have the same solution set?

b. Why should the equations $(x - 1)(x + 3) = 17 + x$ and $(x + 3)(x - 1) = 17 + x$ have the same solution set?

c. Do you think the equations $(x - 1)(x + 3) = 17 + x$ and $(x - 1)(x + 3) + 500 = 517 + x$ should have the same solution set? Why?

d. Do you think the equations $(x - 1)(x + 3) = 17 + x$ and $3(x - 1)(x + 3) = 51 + 3x$ should have the same solution set? Explain why.

Exercise 1

a. Use the commutative property to write an equation that has the same solution set as
$$x^2 - 3x + 4 = (x + 7)(x - 12)(5).$$

b. Use the associative property to write an equation that has the same solution set as
$$x^2 - 3x + 4 = (x + 7)(x - 12)(5).$$

c. Does this reasoning apply to the distributive property as well?

Exercise 2

Consider the equation $x^2 + 1 = 7 - x$.

a. Verify that this has the solution set $\{-3, 2\}$. Draw this solution set as a graph on the number line. We will later learn how to show that these happen to be the ONLY solutions to this equation.

b. Let's add 4 to both sides of the equation and consider the new equation $x^2 + 5 = 11 - x$. Verify 2 and -3 are still solutions.

c. Let's now add x to both sides of the equation and consider the new equation $x^2 + 5 + x = 11$. Are 2 and -3 still solutions?

d. Let's add -5 to both sides of the equation and consider the new equation $x^2 + x = 6$. Are 2 and -3 still solutions?

e. Let's multiply both sides by $\frac{1}{6}$ to get $\frac{x^2+x}{6} = 1$. Are 2 and -3 still solutions?

f. Let's go back to part (d) and add $3x^3$ to both sides of the equation and consider the new equation $x^2 + x + 3x^3 = 6 + 3x^3$. Are 2 and -3 still solutions?

EUREKA
MATH™

Exercise 3

a.   Solve for r: $\dfrac{3}{2r} = \dfrac{1}{4}$

b.   Solve for s: $s^2 + 5 = 30$

c.   Solve for y: $4y - 3 = 5y - 8$

Exercise 4

Consider the equation $3x + 4 = 8x - 16$. Solve for x using the given starting point.

Group 1	Group 2	Group 3	Group 4
Subtract $3x$ from both sides	Subtract 4 from both sides	Subtract $8x$ from both sides	Add 16 to both sides

Closing

Consider the equation $3x^2 + x = (x - 2)(x + 5)x$.

a. Use the commutative property to create an equation with the same solution set.

b. Using the result from part (a), use the associative property to create an equation with the same solution set.

c. Using the result from part (b), use the distributive property to create an equation with the same solution set.

d. Using the result from part (c), add a number to both sides of the equation.

e. Using the result from part (d), subtract a number from both sides of the equation.

f. Using the result from part (e), multiply both sides of the equation by a number.

g. Using the result from part (f), divide both sides of the equation by a number.

h. What do all seven equations have in common? Justify your answer.

Lesson 12:   Solving Equations

**EUREKA
MATH**

Lesson Summary

If x is a solution to an equation, it will also be a solution to the new equation formed when the same number is added to (or subtracted from) each side of the original equation or when the two sides of the original equation are multiplied by (or divided by) the same nonzero number. These are referred to as the *properties of equality*.

If one is faced with the task of solving an equation, that is, finding the solution set of the equation:

Use the *commutative*, *associative*, and *distributive properties*, AND use the *properties of equality* (adding, subtracting, multiplying by nonzeros, dividing by nonzeros) to keep rewriting the equation into one whose solution set you easily recognize. (We believe that the solution set will not change under these operations.)

Problem Set

1. Which of the following equations have the same solution set? Give reasons for your answers that do not depend on solving the equations.

   I. $x - 5 = 3x + 7$

   II. $3x - 6 = 7x + 8$

   III. $15x - 9 = 6x + 24$

   IV. $6x - 16 = 14x + 12$

   V. $9x + 21 = 3x - 15$

   VI. $-0.05 + \frac{x}{100} = \frac{3x}{100} + 0.07$

Solve the following equations, check your solutions, and then graph the solution sets.

2. $-16 - 6v = -2(8v - 7)$

3. $2(6b + 8) = 4 + 6b$

4. $x^2 - 4x + 4 = 0$

5. $7 - 8x = 7(1 + 7x)$

6. $39 - 8n = -8(3 + 4n) + 3n$

7. $(x - 1)(x + 5) = x^2 + 4x - 2$

8. $x^2 - 7 = x^2 - 6x - 7$

9. $-7 - 6a + 5a = 3a - 5a$

10. $7 - 2x = 1 - 5x + 2x$

11. $4(x - 2) = 8(x - 3) - 12$

12. $-3(1 - n) = -6 - 6n$

13. $-21 - 8a = -5(a + 6)$

14. $-11 - 2p = 6p + 5(p + 3)$

15. $\frac{x}{x+2} = 4$

16. $2 + \frac{x}{9} = \frac{x}{3} - 3$

17. $-5(-5x - 6) = -22 - x$

18. $\frac{x+4}{3} = \frac{x+2}{5}$

19. $-5(2r - 0.3) + 0.5(4r + 3) = -64$

EUREKA MATH™

This page intentionally left blank

Lesson 13: Some Potential Dangers When Solving Equations

In previous lessons, we have looked at techniques for solving equations, a common theme throughout algebra. In this lesson, we examine some potential dangers where our intuition about algebra may need to be examined.

Classwork

Exercises

1. Describe the property used to convert the equation from one line to the next:

$$x(1 - x) + 2x - 4 = 8x - 24 - x^2$$
$$x - x^2 + 2x - 4 = 8x - 24 - x^2 \quad \rule{4cm}{0.4pt}$$
$$x + 2x - 4 = 8x - 24 \quad \rule{4cm}{0.4pt}$$
$$3x - 4 = 8x - 24 \quad \rule{4cm}{0.4pt}$$
$$3x + 20 = 8x \quad \rule{4cm}{0.4pt}$$
$$20 = 5x \quad \rule{4cm}{0.4pt}$$

In each of the steps above, we applied a property of real numbers and/or equations to create a new equation.

a. Why are we sure that the initial equation $x(1 - x) + 2x - 4 = 8x - 24 - x^2$ and the final equation $20 = 5x$ have the same solution set?

b. What is the common solution set to all these equations?

2. Solve the equation for x. For each step, describe the operation used to convert the equation.

$$3x - [8 - 3(x - 1)] = x + 19$$

3. Solve each equation for x. For each step, describe the operation used to convert the equation.

   a. $7x - [4x - 3(x - 1)] = x + 12$

   b. $2[2(3 - 5x) + 4] = 5[2(3 - 3x) + 2]$

   c. $\frac{1}{2}(18 - 5x) = \frac{1}{3}(6 - 4x)$

EUREKA
MATH™

4. Consider the equations $x + 1 = 4$ and $(x + 1)^2 = 16$.

   a. Verify that $x = 3$ is a solution to both equations.

   b. Find a second solution to the second equation.

   c. Based on your results, what effect does squaring both sides of an equation appear to have on the solution set?

5. Consider the equations $x - 2 = 6 - x$ and $(x - 2)^2 = (6 - x)^2$.

   a. Did squaring both sides of the equation affect the solution sets?

   b. Based on your results, does your answer to part (c) of the previous question need to be modified?

6. Consider the equation $x^3 + 2 = 2x^2 + x$.

    a. Verify that $x = 1$, $x = -1$, and $x = 2$ are each solutions to this equation.

    b. Bonzo decides to apply the action "ignore the exponents" on each side of the equation. He gets $x + 2 = 2x + x$. In solving this equation, what does he obtain? What seems to be the problem with his technique?

    c. What would Bonzo obtain if he applied his "method" to the equation $x^2 + 4x + 2 = x^4$? Is it a solution to the original equation?

7. Consider the equation $x - 3 = 5$.

    a. Multiply both sides of the equation by a constant, and show that the solution set did not change.

        Now, multiply both sides by x.

    b. Show that $x = 8$ is still a solution to the new equation.

Lesson 13:   Some Potential Dangers When Solving Equations

EUREKA
MATH™

c.   Show that $x = 0$ is also a solution to the new equation.

Now, multiply both sides by the factor $x - 1$.

d.   Show that $x = 8$ is still a solution to the new equation.

e.   Show that $x = 1$ is also a solution to the new equation.

f.   Based on your results, what effect does multiplying both sides of an equation by a constant have on the solution set of the new equation?

g.   Based on your results, what effect does multiplying both sides of an equation by a variable factor have on the solution set of the new equation?

Lesson Summary

Assuming that there is a solution to an equation, applying the distributive, commutative, and associative properties and the properties of equality to equations will not change the solution set.

Feel free to try doing other operations to both sides of an equation, but be aware that the new solution set you get contains possible <u>candidates</u> for solutions. You have to plug each one into the original equation to see if it really is a solution to your original equation.

Problem Set

1. Solve each equation for x. For each step, describe the operation used to convert the equation. How do you know that the initial equation and the final equation have the same solution set?

   a. $\frac{1}{5}[10 - 5(x - 2)] = \frac{1}{10}(x + 1)$

   b. $x(5 + x) = x^2 + 3x + 1$

   c. $2x(x^2 - 2) + 7x = 9x + 2x^3$

2. Consider the equation $x + 1 = 2$.

   a. Find the solution set.

   b. Multiply both sides by $x + 1$, and find the solution set of the new equation.

   c. Multiply both sides of the original equation by x, and find the solution set of the new equation.

3. Solve the equation $x + 1 = 2x$ for x. Square both sides of the equation, and verify that your solution satisfies this new equation. Show that $-\frac{1}{3}$ satisfies the new equation but not the original equation.

4. Consider the equation $x^3 = 27$.

   a. What is the solution set?

   b. Does multiplying both sides by x change the solution set?

   c. Does multiplying both sides by x^2 change the solution set?

5. Consider the equation $x^4 = 16$.

   a. What is the solution set?

   b. Does multiplying both sides by x change the solution set?

   c. Does multiplying both sides by x^2 change the solution set?

Lesson 14: Solving Inequalities

Classwork

Exercise 1

1. Consider the inequality $x^2 + 4x \geq 5$.

    a. Sift through some possible values to assign to x that make this inequality a true statement. Find at least two positive values that work and at least two negative values that work.

    b. Should your four values also be solutions to the inequality $x(x + 4) \geq 5$? Explain why or why not. Are they?

    c. Should your four values also be solutions to the inequality $4x + x^2 \geq 5$? Explain why or why not. Are they?

    d. Should your four values also be solutions to the inequality $4x + x^2 - 6 \geq -1$? Explain why or why not. Are they?

    e. Should your four values also be solutions to the inequality $12x + 3x^2 \geq 15$? Explain why or why not. Are they?

Example 1

What is the solution set to the inequality $5q + 10 > 20$? Express the solution set in words, in set notation, and graphically on the number line.

Exercises 2–3

2. Find the solution set to each inequality. Express the solution in set notation and graphically on the number line.

   a. $x + 4 \leq 7$

   b. $\dfrac{m}{3} + 8 \neq 9$

   c. $8y + 4 < 7y - 2$

   d. $6(x - 5) \geq 30$

   e. $4(x - 3) > 2(x - 2)$

Lesson 14:   Solving Inequalities

EUREKA
MATH™

3. Recall the discussion on all the strange ideas for what could be done to both sides of an equation. Let's explore some of the same issues here but with inequalities. Recall, in this lesson, we have established that adding (or subtracting) and multiplying through by positive quantities does not change the solution set of an inequality. We've made no comment about other operations.

a. Squaring: Do $B \le 6$ and $B^2 \le 36$ have the same solution set? If not, give an example of a number that is in one solution set but not the other.

b. Multiplying through by a negative number: Do $5 - C > 2$ and $-5 + C > -2$ have the same solution set? If not, give an example of a number that is in one solution set but not the other.

c. Bonzo's ignoring exponents: Do $y^2 < 5^2$ and $y < 5$ have the same solution set?

Example 2

Jojo was asked to solve $6x + 12 < 3x + 6$, for x. She answered as follows:

$6x + 12 < 3x + 6$

$6(x + 2) < 3(x + 2)$ Apply the distributive property.

$6 < 3$ Multiply through by $\dfrac{1}{x+2}$.

a. Since the final line is a false statement, she deduced that there is no solution to this inequality (that the solution set is empty).

What is the solution set to $6x + 12 < 3x + 6$?

b. Explain why Jojo came to an erroneous conclusion.

Example 3

Solve $-q \geq -7$, for q.

EUREKA
MATH™

Exercises 4–7

4. Find the solution set to each inequality. Express the solution in set notation and graphically on the number line.

a. $-2f < -16$

b. $-\frac{x}{12} \leq \frac{1}{4}$

c. $6 - a \geq 15$

d. $-3(2x + 4) > 0$

Recall the properties of inequality:

-   Addition property of inequality:

    If $A > B$, then $A + c > B + c$ for any real number c.

-   Multiplication property of inequality:

    If $A > B$, then $kA > kB$ for any <u>positive</u> real number k.

---

5. Use the properties of inequality to show that each of the following is true for any real numbers p and q.

a.   If $p \geq q$, then $-p \leq -q$.

b.   If $p < q$, then $-5p > -5q$.

c. If $p \le q$, then $-0.03p \ge -0.03q$.

d. Based on the results from parts (a) through (c), how might we expand the multiplication property of inequality?

6. Solve $-4 + 2t - 14 - 18t > -6 - 100t$, for t in two different ways: first without ever multiplying through by a negative number and then by first multiplying through by $-\frac{1}{2}$.

7. Solve $-\frac{x}{4} + 8 < \frac{1}{2}$, for x in two different ways: first without ever multiplying through by a negative number and then by first multiplying through by -4.

Lesson 14:   Solving Inequalities

EUREKA MATH™

Problem Set

1. Find the solution set to each inequality. Express the solution in set notation and graphically on the number line.

   a. $2x < 10$ $x < 5$

   b. $-15x \geq -45$ $x \leq 3$

   c. $\frac{2}{3}x \neq \frac{1}{2} + 2$

   d. $-5(x - 1) \geq 10$ $-5x + 5 \geq 10$ $x \leq -1$

   e. $13x < 9(1 - x)$ $13x < 9 - 9x$ $22x < \frac{9}{22}$ $x < \frac{9}{22}$

2. Find the mistake in the following set of steps in a student's attempt to solve $5x + 2 \geq x + \frac{2}{5}$, for x. What is the correct solution set?

   $$5x + 2 \geq x + \frac{2}{5}$$

   $$5\left(x + \frac{2}{5}\right) \geq x + \frac{2}{5} \quad \text{(factoring out 5 on the left side)}$$

   $$5 \geq 1 \quad \text{(dividing by } \left(x + \frac{2}{5}\right))$$

   So, the solution set is the empty set.

3. Solve $-\frac{x}{16} + 1 \geq -\frac{5x}{2}$, for x without multiplying by a negative number. Then, solve by multiplying through by -16.

4. Lisa brought half of her savings to the bakery and bought 12 croissants for $14.20. The amount of money she brings home with her is more than $2.00. Use an inequality to find how much money she had in her savings before going to the bakery. (Write the inequality that represents the situation, and solve it.)

This page intentionally left blank

Lesson 15: Solution Sets of Two or More Equations (or Inequalities) Joined by "And" or "Or"

Classwork

Exercise 1

Determine whether each claim given below is true or false.

a. Right now, I am in math class and English class.

b. Right now, I am in math class or English class.

c. $3 + 5 = 8$ and $5 < 7 - 1$

d. $10 + 2 \neq 12$ and $8 - 3 > 0$

e. $3 < 5 + 4$ or $6 + 4 = 9$

f. $16 - 20 > 1$ or $5.5 + 4.5 = 11$

These are all examples of declarative compound sentences.

g. When the two declarations in the sentences above were separated by "and," what had to be true to make the statement true?

h. When the two declarations in the sentences above were separated by "or," what had to be true to make the statement true?

Example 1

Solve each system of equations and inequalities.

a. $x + 8 = 3$ or $x - 6 = 2$

b. $4x - 9 = 0$ or $3x + 5 = 2$

c. $x - 6 = 1$ and $x + 2 = 9$

d. $2w - 8 = 10$ and $w > 9$

Exercise 2

a. Using a colored pencil, graph the inequality $x < 3$ on the number line below part (c).

b. Using a different colored pencil, graph the inequality $x > -1$ on the same number line.

c. Using a third colored pencil, darken the section of the number line where $x < 3$ and $x > -1$.

d. Using a colored pencil, graph the inequality $x < -4$ on the number line below part (f).

e. Using a different colored pencil, graph the inequality $x > 0$ on the same number line.

f. Using a third colored pencil, darken the section of the number line where $x < -4$ or $x > 0$.

g.   Graph the compound sentence $x > -2$ or $x = -2$ on the number line below.

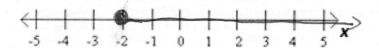

h.   How could we abbreviate the sentence $x > -2$ or $x = -2$?

$$x \geq -2$$

i.   Rewrite $x \leq 4$ as a compound sentence, and graph the solutions to the sentence on the number line below.

Example 2

Graph each compound sentence on a number line.

a. $x = 2$ or $x > 6$

b. $x \leq -5$ or $x \geq 2$

Rewrite as a compound sentence, and graph the sentence on a number line.

c. $1 \leq x \leq 3$

Exercise 3

Consider the following two scenarios. For each, specify the variable and say, "W is the width of the rectangle," for example, and write a compound inequality that represents the scenario given. Draw its solution set on a number line.

Scenario	Variable	Inequality	Graph
a. Students are to present a persuasive speech in English class. The guidelines state that the speech must be at least 7 minutes but not exceed 12 minutes.			
b. Children and senior citizens receive a discount on tickets at the movie theater. To receive a discount, a person must be between the ages of 2 and 12, including 2 and 12, or 60 years of age or older.			

Exercise 4

Determine if each sentence is true or false. Explain your reasoning.

a. $8 + 6 \leq 14$ and $\dfrac{1}{3} < \dfrac{1}{2}$

b. $5 - 8 < 0$ or $10 + 13 \neq 23$

Solve each system, and graph the solution on a number line.

c. $x - 9 = 0$ or $x + 15 = 0$

d. $5x - 8 = -23$ or $x + 1 = -10$

EUREKA
MATH™

Graph the solution set to each compound inequality on a number line.

e. $x < -8$ or $x > -8$

f. $0 < x \leq 10$

Write a compound inequality for each graph.

g.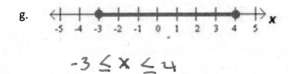

$-3 \leq x \leq 4$

h.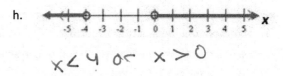

$x < 4$ or $x > 0$

i. A poll shows that a candidate is projected to receive 57% of the votes. If the margin for error is plus or minus 3%, write a compound inequality for the percentage of votes the candidate can expect to get.

j. Mercury is one of only two elements that are liquid at room temperature. Mercury is nonliquid for temperatures less than $-38.0°F$ or greater than $673.8°F$. Write a compound inequality for the temperatures at which mercury is nonliquid.

Lesson Summary

In mathematical sentences, like in English sentences, a compound sentence separated by

AND is true if _____.

OR is true if _____.

Problem Set

1. Consider the inequality $0 < x < 3$.
   a. Rewrite the inequality as a compound sentence.
   b. Graph the inequality on a number line.
   c. How many solutions are there to the inequality? Explain.
   d. What are the largest and smallest possible values for x? Explain.
   e. If the inequality is changed to $0 \le x \le 3$, then what are the largest and smallest possible values for x?

Write a compound inequality for each graph.

2.
                                                                                      3.

Write a single or compound inequality for each scenario.
4. The scores on the last test ranged from 65% to 100%.

5. To ride the roller coaster, one must be at least 4 feet tall.

6. Unsafe body temperatures are those lower than 96°F or above 104°F.

Graph the solution(s) to each of the following on a number line.

7. $x - 4 = 0$ and $3x + 6 = 18$ 8. $x < 5$ and $x \ne 0$

9. $x \le -8$ or $x \ge -1$ 10. $3(x - 6) = 3$ or $5 - x = 2$

11. $x < 9$ and $x > 7$ 12. $x + 5 < 7$ or $x = 2$

EUREKA
MATH™

Lesson 16: Solving and Graphing Inequalities Joined by "And" or "Or"

Classwork

Exercise 1

a. Solve $w^2 = 121$, for w. Graph the solution on a number line.

b. Solve $w^2 < 121$, for w. Graph the solution on a number line, and write the solution set as a compound inequality.

c. Solve $w^2 \geq 121$, for w. Graph the solution on a number line, and write the solution set as a compound inequality.

d. Quickly solve $(x + 7)^2 = 121$, for x. Graph the solution on a number line.

e. Use your work from part (d) to quickly graph the solution on a number line to each inequality below.
    i. $(x + 7)^2 < 121$

    ii. $(x + 7)^2 \geq 121$

Exercise 2

Consider the compound inequality $-5 < x < 4$.

    a.   Rewrite the inequality as a compound statement of inequality.

    b.   Write a sentence describing the possible values of x.

    c.   Graph the solution set on the number line below.

Exercise 3

Consider the compound inequality $-5 < 2x + 1 < 4$.

    a.   Rewrite the inequality as a compound statement of inequality.

    b.   Solve each inequality for x. Then, write the solution to the compound inequality.

    c.   Write a sentence describing the possible values of x.

    d.   Graph the solution set on the number line below.

EUREKA
MATH™

Exercise 4

Given $x < -3$ or $x > -1$:

a. What must be true in order for the compound inequality to be a true statement?

b. Write a sentence describing the possible values of x.

c. Graph the solution set on the number line below.

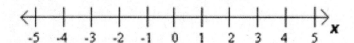

Exercise 5

Given $x + 4 < 6$ or $x - 1 > 3$:

a. Solve each inequality for x. Then, write the solution to the compound inequality.

b. Write a sentence describing the possible values of x.

c. Graph the solution set on the number line below.

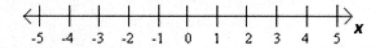

Exercise 6

Solve each compound inequality for x, and graph the solution on a number line.

a. $x + 6 < 8$ and $x - 1 > -1$

b. $-1 \leq 3 - 2x \leq 10$

c. $5x + 1 < 0$ or $8 \leq x - 5$

d. $10 > 3x - 2$ or $x = 4$

e. $x - 2 < 4$ or $x - 2 > 4$

f. $x - 2 \leq 4$ and $x - 2 \geq 4$

EUREKA
MATH™

Exercise 7

Solve each compound inequality for x, and graph the solution on a number line. Pay careful attention to the inequality symbols and the "and" or "or" statements as you work.

a. $1 + x > -4$ or $3x - 6 > -12$

b. $1 + x > -4$ or $3x - 6 < -12$

c. $1 + x > 4$ and $3x - 6 < -12$

Problem Set

Solve each inequality for x, and graph the solution on a number line.

1. $x - 2 < 6$ or $\dfrac{x}{3} > 4$

2. $-6 < \dfrac{x+1}{4} < 3$

3. $5x \le 21 + 2x$ or $3(x + 1) \ge 24$

4. $5x + 2 \ge 27$ and $3x - 1 < 29$

5. $0 \le 4x - 3 \le 11$

6. $2x > 8$ or $-2x < 4$

7. $8 \ge -2(x - 9) \ge -8$

8. $4x + 8 > 2x - 10$ or $\dfrac{1}{3}x - 3 < 2$

9. $7 - 3x < 16$ and $x + 12 < -8$

10. If the inequalities in Problem 8 were joined by "and" instead of "or," what would the solution set become?

11. If the inequalities in Problem 9 were joined by "or" instead of "and," what would the solution set become?

EUREKA
MATH™

Lesson 17: Equations Involving Factored Expressions

Classwork

Exercise 1

1. Solve each equation for x.

   a. $x - 10 = 0$

   b. $\dfrac{x}{2} + 20 = 0$

   c. Demanding Dwight insists that you give him two solutions to the following equation:

   $$(x - 10)\left(\dfrac{x}{2} + 20\right) = 0$$

   Can you provide him with two solutions?

   d. Demanding Dwight now wants FIVE solutions to the following equation:

   $$(x - 10)(2x + 6)(x^2 - 36)(x^2 + 10)\left(\dfrac{x}{2} + 20\right) = 0$$

   Can you provide him with five solutions?

   Do you think there might be a sixth solution?

Consider the equation $(x - 4)(x + 3) = 0$.

e.   Rewrite the equation as a compound statement.

f.   Find the two solutions to the equation.

Example 1

Solve $2x^2 - 10x = 0$, for x.

Example 2

Solve $x(x - 3) + 5(x - 3) = 0$, for x.

Exercises 2–7

2. $(x + 1)(x + 2) = 0$

3. $(3x - 2)(x + 12) = 0$

4. $(x - 3)(x - 3) = 0$

EUREKA
MATH™

5. $(x + 4)(x - 6)(x - 10) = 0$ 6. $x^2 - 6x = 0$ 7. $x(x - 5) + 4(x - 5) = 0$

Example 3

Consider the equation $(x - 2)(2x - 3) = (x - 2)(x + 5)$. Lulu chooses to multiply through by $\frac{1}{x-2}$ and gets the answer $x = 8$. But Poindexter points out that $x = 2$ is also an answer, which Lulu missed.

   a.   What's the problem with Lulu's approach?

   b.   Use factoring to solve the original equation for x.

Exercises 8–11

8. Use factoring to solve the equation for x: $(x-2)(2x-3) = (x-2)(x+1)$.

9. Solve each of the following for x:

    a. $x + 2 = 5$

    b. $x^2 + 2x = 5x$

    c. $x(5x - 20) + 2(5x - 20) = 5(5x - 20)$

10.

    a. Verify: $(a-5)(a+5) = a^2 - 25$.

    b. Verify: $(x-88)(x+88) = x^2 - 88^2$.

Lesson 17:   Equations Involving Factored Expressions

EUREKA MATH™

c.   Verify: $A^2 - B^2 = (A - B)(A + B)$.

d.   Solve for x: $x^2 - 9 = 5(x - 3)$.

e.   Solve for w: $(w + 2)(w - 5) = w^2 - 4$.

11. A string 60 inches long is to be laid out on a tabletop to make a rectangle of perimeter 60 inches. Write the width of the rectangle as $15 + x$ inches. What is an expression for its length? What is an expression for its area? What value for x gives an area of the largest possible value? Describe the shape of the rectangle for this special value of x.

Lesson Summary

The *zero-product property* says that if $ab = 0$, then either $a = 0$ or $b = 0$ or $a = b = 0$.

Problem Set

1. Find the solution set of each equation:
   a. $(x - 1)(x - 2)(x - 3) = 0$
   b. $(x - 16.5)(x - 109) = 0$
   c. $x(x + 7) + 5(x + 7) = 0$
   d. $x^2 + 8x + 15 = 0$
   e. $(x - 3)(x + 3) = 8x$

2. Solve $x^2 - 11x = 0$, for x.

3. Solve $(p + 3)(p - 5) = 2(p + 3)$, for p. What solution do you lose if you simply divide by $p + 3$ to get $p - 5 = 2$?

4. The square of a number plus 3 times the number is equal to 4. What is the number?

5. In the right triangle shown below, the length of side AB is x, the length of side BC is $x + 2$, and the length of the hypotenuse AC is $x + 4$. Use this information to find the length of each side. (Use the Pythagorean theorem to get an equation, and solve for x.)

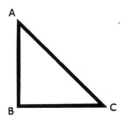

6. Using what you learned in this lesson, create an equation that has 53 and 22 as its only solutions.

Lesson 17:   Equations Involving Factored Expressions

**EUREKA
MATH™**

Lesson 18: Equations Involving a Variable Expression in the Denominator

Classwork

Opening Exercise

Nolan says that he checks the answer to a division problem by performing multiplication. For example, he says that $20 \div 4 = 5$ is correct because 5×4 is 20, and $\frac{3}{\frac{1}{2}} = 6$ is correct because $6 \times \frac{1}{2}$ is 3.

  a. Using Nolan's reasoning, explain why there is no real number that is the answer to the division problem $5 \div 0$.

  b. Quentin says that $\frac{0}{0} = 17$. What do you think?

  c. Mavis says that the expression $\frac{5}{x+2}$ has a meaningful value for whatever value one chooses to assign to x. Do you agree?

  d. Bernoit says that the expression $\frac{3x-6}{x-2}$ always has the value 3 for whichever value one assigns to x. Do you agree?

Exercises 1–2

1. Rewrite $\dfrac{10}{x+5}$ as a compound statement.

2. Consider $\dfrac{x^2-25}{(x^2-9)(x+4)}$.

   a. Is it permissible to let $x = 5$ in this expression?

   b. Is it permissible to let $x = 3$ in this expression?

   c. Give all the values of x that are **not** permissible in this expression.

Example 1

Consider the equation $\dfrac{1}{x} = \dfrac{3}{x-2}$.

   a. Rewrite the equation into a system of equations.

   b. Solve the equation for x, excluding the value(s) of x that lead to a denominator of zero.

Example 2

Consider the equation $\dfrac{x+3}{x-2} = \dfrac{5}{x-2}$.

    a.   Rewrite the equation into a system of equations.

    b.   Solve the equation for x, excluding the value(s) of x that lead to a denominator of zero.

Exercises 3–11

Rewrite each equation into a system of equations excluding the value(s) of x that lead to a denominator of zero; then, solve the equation for x.

3. $\dfrac{5}{x} = 1$

4. $\dfrac{1}{x-5} = 3$

5. $\dfrac{x}{x+1} = 4$

6. $\dfrac{2}{x} = \dfrac{3}{x-4}$

7. $\dfrac{x}{x+6} = -\dfrac{6}{x+6}$

8. $\dfrac{x-3}{x+2} = 0$

9. $\dfrac{x+3}{x+3} = 5$

10. $\dfrac{x+3}{x+3} = 1$

11. A baseball player's batting average is calculated by dividing the number of times a player got a hit by the total number of times the player was at bat. It is expressed as a decimal rounded to three places. After the first 10 games of the season, Samuel had 12 hits off of 33 at bats.

   a. What is his batting average after the first 10 games?

   b. How many hits in a row would he need to get to raise his batting average to above 0.500?

   c. How many *at bats* in a row without a hit would result in his batting average dropping below 0.300?

EUREKA
MATH™

Problem Set

1. Consider the equation $\dfrac{10(x^2-49)}{3x(x^2-4)(x+1)} = 0$. Is $x = 7$ permissible? Which values of x are excluded? Rewrite as a system of equations.

2. Rewrite each equation as a system of equations excluding the value(s) of x that lead to a denominator of zero. Then, solve the equation for x.

    a. $25x = \dfrac{1}{x}$

    b. $\dfrac{1}{5x} = 10$

    c. $\dfrac{x}{7-x} = 2x$

    d. $\dfrac{2}{x} = \dfrac{5}{x+1}$

    e. $\dfrac{3+x}{3-x} = \dfrac{3+2x}{3-2x}$

3. Ross wants to cut a 40-foot rope into two pieces so that the length of the first piece divided by the length of the second piece is 2.

    a. Let x represent the length of the first piece. Write an equation that represents the relationship between the pieces as stated above.

    b. What values of x are not permissible in this equation? Describe within the context of the problem what situation is occurring if x were to equal this value(s). Rewrite as a system of equations to exclude the value(s).

    c. Solve the equation to obtain the lengths of the two pieces of rope. (Round to the nearest tenth if necessary.)

4. Write an equation with the restrictions $x \neq 14$, $x \neq 2$, and $x \neq 0$.

5. Write an equation that has no solution.

This page intentionally left blank

Lesson 19: Rearranging Formulas

Classwork

Exercises 1–3

1. Solve each equation for x. For part (c), remember a variable symbol, like a, b, and c, represents a number.

   a. $2x - 6 = 10$

   b. $-3x - 3 = -12$

   c. $ax - b = c$

2. Compare your work in parts (a) through (c) above. Did you have to do anything differently to solve for x in part (c)?

3. Solve the equation $ax - b = c$ for a. The variable symbols x, b, and c, represent numbers.

Example 1: Rearranging Familiar Formulas

The area A of a rectangle is 25 in^2. The formula for area is $A = lw$.

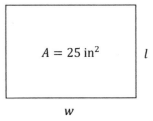

- If the width w is 10 inches, what is the length l?

- If the width w is 15 inches, what is the length l?

- Rearrange the area formula to solve for l.

$$A = lw$$

$$\frac{A}{w} = \frac{lw}{w}$$

- Verify that the area formula, solved for l, will give the same results for l as having solved for l in the original area formula.

Exercises 4–5

4. Solve each problem two ways. First, substitute the given values, and solve for the given variable. Then, solve for the given variable, and substitute the given values.

   a. The perimeter formula for a rectangle is $p = 2(l + w)$, where p represents the perimeter, l represents the length, and w represents the width. Calculate l when $p = 70$ and $w = 15$.

   b. The area formula for a triangle is $A = \frac{1}{2}bh$, where A represents the area, b represents the length of the base, and h represents the height. Calculate b when $A = 100$ and $h = 20$.

5. Rearrange each formula to solve for the specified variable. Assume no variable is equal to 0.

    a. Given $A = P(1 + rt)$,

        i. Solve for P.

        ii. Solve for t.

    b. Given $K = \frac{1}{2}mv^2$,

        i. Solve for m.

        ii. Solve for v.

Example 2: Comparing Equations with One Variable to Those with More Than One Variable

Equation Containing More Than One Variable	Related Equation
Solve $ax + b = d - cx$ for x.	Solve $3x + 4 = 6 - 5x$ for x.
Solve for x. $$\dfrac{ax}{b} + \dfrac{cx}{d} = e$$	Solve for x. $$\dfrac{2x}{5} + \dfrac{x}{7} = 3$$

EUREKA
MATH™

Lesson Summary

The properties and reasoning used to solve equations apply regardless of how many variables appear in an equation or formula. Rearranging formulas to solve for a specific variable can be useful when solving applied problems.

Problem Set

For Problems 1–8, solve for x.

1. $ax + 3b = 2f$

2. $rx + h = sx - k$

3. $3px = 2q(r - 5x)$

4. $\dfrac{x+b}{4} = c$

5. $\dfrac{x}{5} - 7 = 2q$

6. $\dfrac{x}{6} - \dfrac{x}{7} = ab$

7. $\dfrac{x}{m} - \dfrac{x}{n} = \dfrac{1}{p}$

8. $\dfrac{3ax+2b}{c} = 4d$

9. Solve for m.
$$t = \frac{ms}{m+n}$$

10. Solve for u.
$$\frac{1}{u} + \frac{1}{v} = \frac{1}{f}$$

11. Solve for s.
$$A = s^2$$

12. Solve for h.
$$V = \pi r^2 h$$

13. Solve for m.
$$T = 4\sqrt{m}$$

14. Solve for d.
$$F = G\frac{mn}{d^2}$$

15. Solve for y.
$$ax + by = c$$

16. Solve for b_1.
$$A = \frac{1}{2}h(b_1 + b_2)$$

17. The science teacher wrote three equations on a board that relate velocity, v, distance traveled, d, and the time to travel the distance, t, on the board.

$$v = \frac{d}{t} \qquad\qquad t = \frac{d}{v} \qquad\qquad d = vt$$

Would you need to memorize all three equations, or could you just memorize one? Explain your reasoning.

This page intentionally left blank

Lesson 20: Solution Sets to Equations with Two Variables

Classwork

Exercises

1.

    a. Circle all the ordered pairs (x, y) that are solutions to the equation $4x - y = 10$.

           $(3,2)$ $(2,3)$ $(-1,-14)$ $(0,0)$ $(1,-6)$

           $(5,10)$ $(0,-10)$ $(3,4)$ $(6,0)$ $(4,-1)$

    b. How did you decide whether or not an ordered pair was a solution to the equation?

2.

    a. Discover as many additional solutions to the equation $4x - y = 10$ as possible. Consider the best way to organize all the solutions you have found. Be prepared to share the strategies you used to find your solutions.

    b. Now, find five more solutions where one or more variables are negative numbers or non-integer values. Be prepared to share the strategies you used to find your solutions.

    c. How many ordered pairs (x, y) will be in the solution set of the equation $4x - y = 10$?

d. Create a visual representation of the solution set by plotting each solution as a point (x, y) in the coordinate plane.

e. Why does it make sense to represent the solution to the equation $4x - y = 10$ as a line in the coordinate plane?

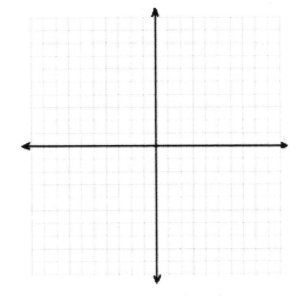

3. The sum of two numbers is 25. What are the numbers?

a. Create an equation using two variables to represent this situation. Be sure to explain the meaning of each variable.

b. List at least six solutions to the equation you created in part (a).

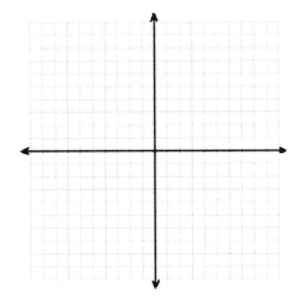

c. Create a graph that represents the solution set to the equation.

Lesson 20:   Solution Sets to Equations with Two Variables

EUREKA MATH™

4. Gia had 25 songs in a playlist composed of songs from her two favorite artists, Beyonce and Jennifer Lopez. How many songs did she have by each one in the playlist?

   a. Create an equation using two variables to represent this situation. Be sure to explain the meaning of each variable.

   b. List at least three solutions to the equation you created in part (a).

   c. Create a graph that represents the solution set to the equation.

5. Compare your solutions to Exercises 3 and 4. How are they alike? How are they different?

Lesson Summary

An *ordered pair* is a *solution* to a two-variable equation when each number substituted into its corresponding variable makes the equation a true number sentence. All of the solutions to a two-variable equation are called the *solution set*.

Each ordered pair of numbers in the solution set of the equation corresponds to a point on the coordinate plane. The set of all such points in the coordinate plane is called the *graph of the equation*.

Problem Set

1. Match each equation with its graph.
   Explain your reasoning.

   a. $y = 5x - 6$

   b. $x + 2y = -12$

   c. $2x + y = 4$

   d. $y = 3x - 6$

   e. $x = -y - 4$

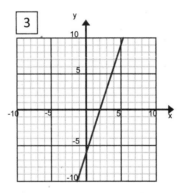

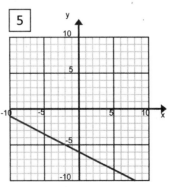

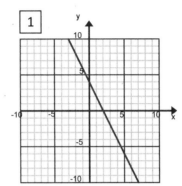

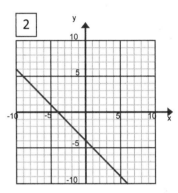

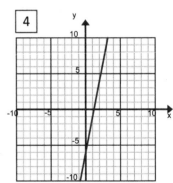

EUREKA
MATH™

2. Graph the solution set in the coordinate plane. Label at least two ordered pairs that are solutions on your graph.

   a. $10x + 6y = 100$ b. $y = 9.5x + 20$ c. $7x - 3y = 21$ d. $y = 4(x + 10)$

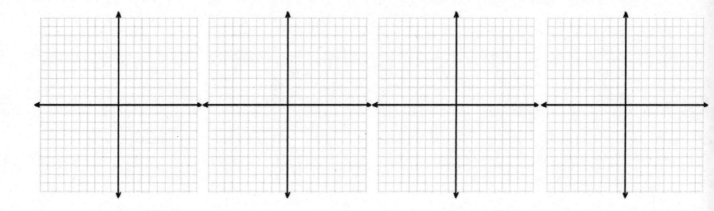

3. Mari and Lori are starting a business to make gourmet toffee. They gather the following information from another business about prices for different amounts of toffee. Which equation and which graph are most likely to model the price, p, for x pounds of toffee? Justify your reasoning.

Pounds, x	Price, p, for x pounds
0.25	$3.60
0.81	$6.48
1	$7.20
1.44	$8.64

Graph 1

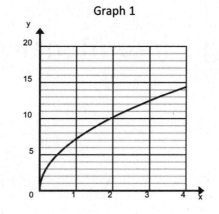

Graph 2

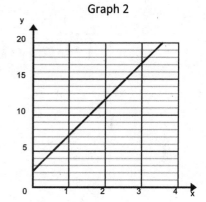

Equation A: $p = 5x + 2.2$

Equation B: $p = 7.2\sqrt{x}$

This page intentionally left blank

Lesson 21: Solution Sets to Inequalities with Two Variables

Classwork

Exercises 1–2

1.

    a.   Circle each ordered pair (x, y) that is a solution to the equation $4x - y \leq 10$.

        i. $(3,2)$ $(2,3)$ $(-1,-14)$ $(0,0)$ $(1,-6)$

        ii. $(5,10)$ $(0,-10)$ $(3,4)$ $(6,0)$ $(4,-1)$

    b.   Plot each solution as a point (x, y) in the coordinate plane.

    c.   How would you describe the location of the solutions in the coordinate plane?

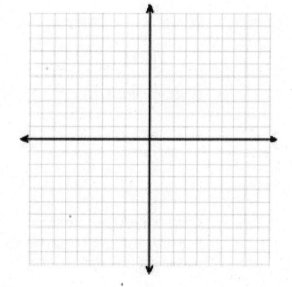

2.

    a.   Discover as many additional solutions to the inequality $4x - y \leq 10$ as possible. Organize solutions by plotting each solution as a point (x, y) in the coordinate plane. Be prepared to share the strategies used to find the solutions.

    b.   Graph the line $y = 4x - 10$. What do we notice about the solutions to the inequality $4x - y \leq 10$ and the graph of the line $y = 4x - 10$?

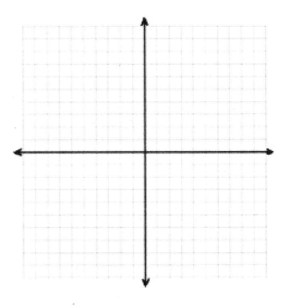

    c.   Solve the inequality for y.

    d.   Complete the following sentence:

        If an ordered pair is a solution to $4x - y \leq 10$, then it will be located _____ the line $y = 4x - 10$.

    e.   Explain how you arrived at your conclusion.

Example

The solution to $x + y = 20$ is shown on the graph below.

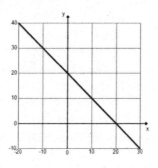

a. Graph the solution to $x + y \leq 20$.

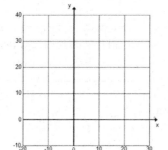

c. Graph the solution to $x + y < 20$.

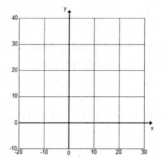

b. Graph the solution to $x + y \geq 20$.

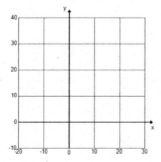

d. Graph the solution to $x + y > 20$.

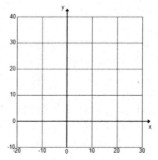

Exercises 3–5

3. Using a separate sheet of graph paper, plot the solution sets to the following equations and inequalities:

    a. $x - y = 10$

    b. $x - y < 10$

    c. $y > x - 10$

    d. $y \geq x$

    e. $x \geq y$

    f. $y = 5$

    g. $y < 5$

    h. $x \geq 5$

    i. $y \neq 1$

    j. $x = 0$

    k. $x > 0$

    l. $y < 0$

    m. $x^2 - y = 0$

    n. $x^2 + y^2 > 0$

    o. $xy \leq 0$

    Which of the inequalities in this exercise are *linear* inequalities?

A *half-plane* is the graph of a solution set in the Cartesian coordinate plane of an inequality in two real-number variables that is linear and strict.

---

4. Describe in words the half-plane that is the solution to each inequality.

    a. $y \geq 0$

    b. $x < -5$

    c. $y \geq 2x - 5$

    d. $y < 2x - 5$

EUREKA MATH™

5. Graph the solution set to $x < -5$, reading it as an inequality in *one* variable, and describe the solution set in words. Then graph the solution set to $x < -5$ again, this time reading it as an inequality in *two* variables, and describe the solution set in words.

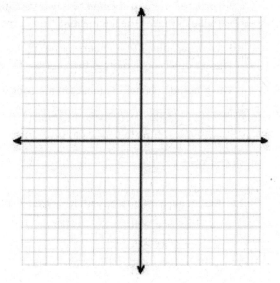

Lesson Summary

An ordered pair is a *solution* to a two-variable inequality if, when each number is substituted into its corresponding variable, it makes the inequality a true number sentence.

Each ordered pair of numbers in the solution set of the inequality corresponds to a point on the coordinate plane. The set of all such points in the coordinate plane is called the *graph of the inequality.*

The graph of a linear inequality in the coordinate plane is called a *half-plane.*

Problem Set

1. Match each inequality with its graph. Explain your reasoning.

    a. $2x - y > 6$

    b. $y \leq 2x - 6$

    c. $2x < y + 6$

    d. $2x - 6 \leq y$

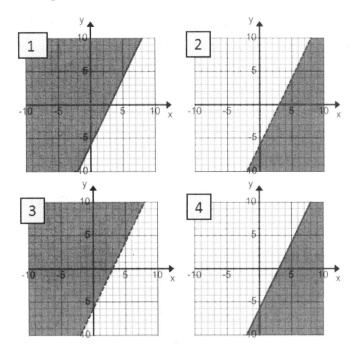

EUREKA
MATH™

2. Graph the solution set in the coordinate plane. Support your answer by selecting two ordered pairs in the solution set and verifying that they make the inequality true.

    a. $-10x + y > 25$

    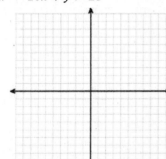

    b. $-6 \leq y$

    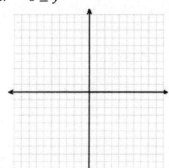

    c. $y \leq -7.5x + 15$

    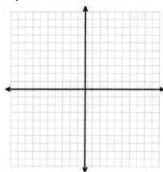

    d. $2x - 8y \leq 24$

    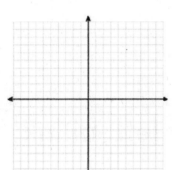

    e. $3x < y$

    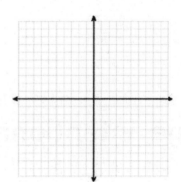

    f. $2x > 0$

    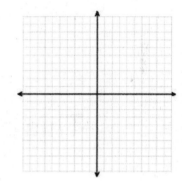

3. Marti sells tacos and burritos from a food truck at the farmers market. She sells burritos for $3.50 each and tacos for $2.00 each. She hopes to earn at least $120 at the farmers market this Saturday.

    a. Identify three combinations of tacos and burritos that will earn Marti more than $120.

    b. Identify three combinations of tacos and burritos that will earn Marti exactly $120.

    c. Identify three combinations of tacos and burritos that will *not* earn Marti at least $120.

    d. Graph your answers to parts (a)–(c) in the coordinate plane, and then shade a half-plane that contains all possible solutions to this problem.

    e. Create a linear inequality that represents the solution to this problem. Let x equal the number of burritos that Marti sells, and let y equal the number of tacos that Marti sells.

    f. Is the point $(10, 49.5)$ a solution to the inequality you created in part (e)? Explain your reasoning.

This page intentionally left blank

Lesson 22: Solution Sets to Simultaneous Equations

Classwork

Opening Exercise

Consider the following compound sentence: $x + y > 10$ and $y = 2x + 1$.

a.   Circle all the ordered pairs (x, y) that are solutions to the inequality $x + y > 10$ (below).

b.   Underline all the ordered pairs (x, y) that are solutions to the equation $y = 2x + 1$.

$(3,7)$	$(7,3)$	$(-1,14)$	$(0,1)$	$(12,25)$
$(5,11)$	$(0,12)$	$(1,8)$	$(12,0)$	$(-1,-1)$

c.   List the ordered pair(s) (x, y) from above that are solutions to the compound sentence $x + y > 10$ and $y = 2x + 1$.

d.   List three additional ordered pairs that are solutions to the compound sentence $x + y > 10$ and $y = 2x + 1$.

e.   Sketch the solution set to the inequality $x + y > 10$ and the solution set to $y = 2x + 1$ on the same set of coordinate axes. Highlight the points that lie in BOTH solution sets.

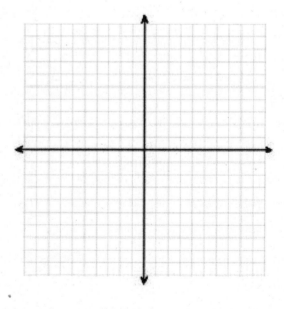

f.   Describe the solution set to $x + y > 10$ and $y = 2x + 1$.

Example 1

Solve the following system of equations.

$$\begin{cases} y = 2x + 1 \\ x - y = 7 \end{cases}$$

Graphically:

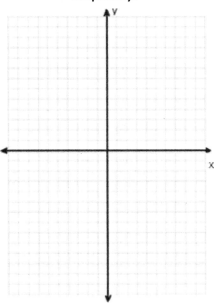

Algebraically:

Exercise 1

Solve each system first by graphing and then algebraically.

a. $\begin{cases} y = 4x - 1 \\ y = -\frac{1}{2}x + 8 \end{cases}$

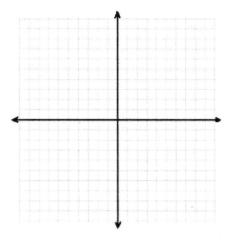

EUREKA
MATH™

b. $\begin{cases} 2x + y = 4 \\ 2x + 3y = 9 \end{cases}$

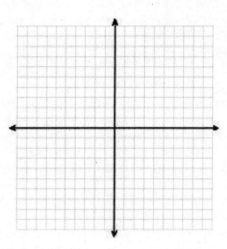

c. $\begin{cases} 3x + y = 5 \\ 3x + y = 8 \end{cases}$

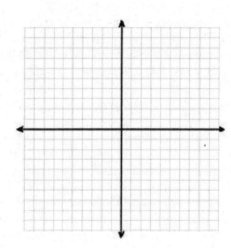

Example 2

Now suppose the system of equations from Exercise 1(c) was instead a system of inequalities:

$\begin{cases} 3x + y \geq 5 \\ 3x + y \leq 8 \end{cases}$

Graph the solution set.

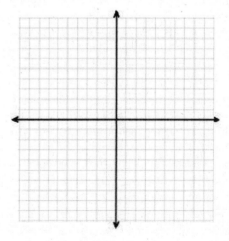

EUREKA MATH™

Example 3

Graph the solution set to the system of inequalities.

$2x - y < 3$ and $4x + 3y \geq 0$

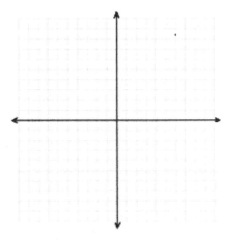

Exercise 2

Graph the solution set to each system of inequalities.

a. $\begin{cases} x - y > 5 \\ x > -1 \end{cases}$

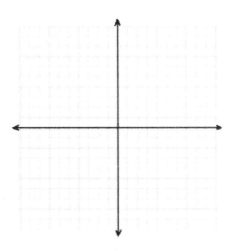

b. $\begin{cases} y \leq x + 4 \\ y \leq 4 - x \\ y \geq 0 \end{cases}$

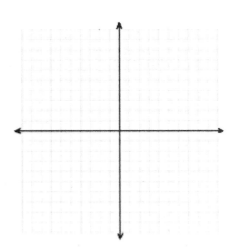

EUREKA
MATH™

Problem Set

1. Estimate the solution to the system of equations by graphing and then find the exact solution to the system algebraically.

$$\begin{cases} 4x + y = -5 \\ x + 4y = 12 \end{cases}$$

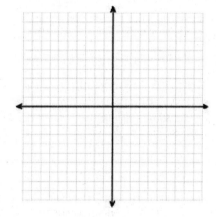

2.

a. Without graphing, construct a system of two linear equations where $(0, 5)$ is a solution to the first equation but is not a solution to the second equation, and $(3, 8)$ is a solution to the system.

b. Graph the system and label the graph to show that the system you created in part (a) satisfies the given conditions.

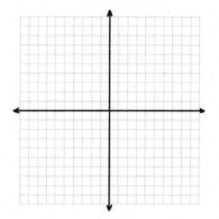

3. Consider two linear equations. The graph of the first equation is shown. A table of values satisfying the second equation is given. What is the solution to the system of the two equations?

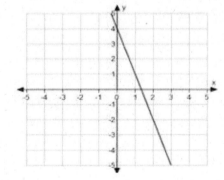

x	-4	-2	0	2	4
y	-26	-18	-10	-2	6

EUREKA
MATH™

4. Graph the solution to the following system of inequalities: $\begin{cases} x \geq 0 \\ y < 2 \\ x + 3y > 0 \end{cases}$

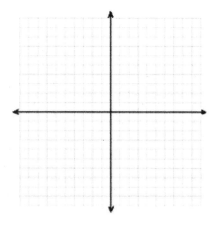

5. Write a system of inequalities that represents the shaded region of the graph shown.

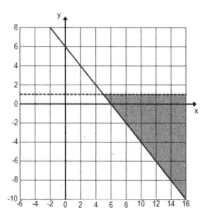

6. For each question below, provide an explanation or an example to support your claim.

   a. Is it possible to have a system of equations that has no solution?

   b. Is it possible to have a system of equations that has more than one solution?

   c. Is it possible to have a system of inequalities that has no solution?

EUREKA
MATH™

Lesson 23: Solution Sets to Simultaneous Equations

Classwork

Opening Exercise

Here is a system of two linear equations. Verify that the solution to this system is (3,4).

> Equation A1: $y = x + 1$
>
> Equation A2: $y = -2x + 10$

Exploratory Challenge

a. Write down another system of two linear equations, B1 and B2, whose solution is $(3, 4)$. This time make sure both linear equations have a positive slope.

b. Verify that the solution to this system of two linear equations is (3,4).

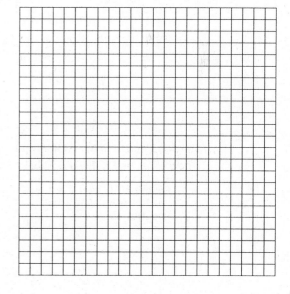

c. Graph equation B1 and B2.

d. Are either B1 or B2 equivalent to the original A1 or A2? Explain your reasoning.

e. Add A1 and A2 to create a new equation C1. Then, multiply A1 by 3 to create a new equation C2. Why is the solution to this system also $(3, 4)$? Explain your reasoning.

The following system of equations was obtained from the original system by adding a multiple of equation A2 to equation A1.

Equation D1: $y = x + 1$

Equation D2: $3y = -3x + 21$

f. What multiple of A2 was added to A1 to create D2?

g. What is the solution to the system of two linear equations formed by D1 and D2?

h. Is D2 equivalent to the original A1 or A2? Explain your reasoning.

i. Start with equation A1. Multiply it by a number of your choice and add the result to equation A2. This creates a new equation E2. Record E2 below to check if the solution is $(3, 4)$.

Equation E1: $y = x + 1$

Equation E2:

EUREKA MATH™

Example: Why Does the Elimination Method Work?

Solve this system of linear equations algebraically.

ORIGINAL SYSTEM	NEW SYSTEM	SOLUTION
$2x + y = 6$		
$x - 3y = -11$		

Exercises

1. Explain a way to create a new system of equations with the same solution as the original that eliminates variable y from one equation. Then determine the solution.

ORIGINAL SYSTEM	NEW SYSTEM	SOLUTION
$2x + 3y = 7$		
$x - y = 1$		

2. Explain a way to create a new system of equations with the same solution as the original that eliminates variable x from one equation. Then determine the solution.

ORIGINAL SYSTEM	NEW SYSTEM	SOLUTION
$2x + 3y = 7$		
$x - y = 1$		

Problem Set

Try to answer the following without solving for x and y first:

1. If $3x + 2y = 6$ and $x + y = 4$, then

   a. $2x + y = ?$ b. $4x + 3y = ?$

2. You always get the same solution no matter which two of the four equations you choose from Problem 1 to form a system of two linear equations. Explain why this is true.

3. Solve the system of equations $\begin{cases} y = \dfrac{1}{4}x \\ y = -x + 5 \end{cases}$ by graphing. Then, create a new system of equations that has the same solution. Show either algebraically or graphically that the systems have the same solution.

4. Without solving the systems, explain why the following systems must have the same solution.

   System (i): $4x - 5y = 13$ System (ii): $8x - 10y = 26$
                  $3x + 6y = 11$ $x - 11y = 2$

Solve each system of equations by writing a new system that eliminates one of the variables.

5. $2x + y = 25$
   $4x + 3y = 9$

6. $3x + 2y = 4$
   $4x + 7y = 1$

Lesson 24: Applications of Systems of Equations and Inequalities

Classwork

Opening Exercise

In Lewis Carroll's *Through the Looking Glass,* Tweedledum says, "The sum of your weight and twice mine is 361 pounds." Tweedledee replies, "The sum of your weight and twice mine is 362 pounds." Find both of their weights.

Example

Lulu tells her little brother, Jack, that she is holding 20 coins, all of which are either dimes or quarters. They have a value of $4.10. She says she will give him the coins if he can tell her how many of each she is holding. Solve this problem for Jack.

Exploratory Challenge

a. At a state fair, there is a game where you throw a ball at a pyramid of cans. If you knock over all of the cans, you win a prize. The cost is 3 throws for $1, but if have you an armband, you get 6 throws for $1. The armband costs $10.

    i. Write two cost equations for the game in terms of the number of throws purchased, one without an armband and one with.

    ii. Graph the two cost equations on the same graph. Be sure to label the axes and show an appropriate scale.

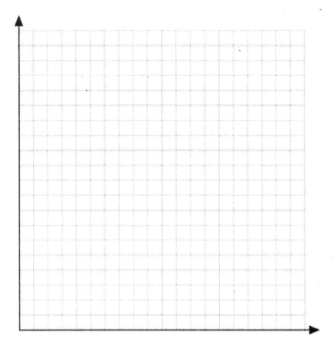

    iii. Does it make sense to buy the armband?

　　Lesson 24:　Applications of Systems of Equations and Inequalities

b.   A clothing manufacturer has 1,000 yd. of cotton to make shirts and pajamas. A shirt requires 1 yd. of fabric, and a pair of pajamas requires 2 yd. of fabric. It takes 2 hr. to make a shirt and 3 hr. to make the pajamas, and there are 1,600 hr. available to make the clothing.

   i.   What are the variables?

   ii.   What are the constraints?

   iii.   Write inequalities for the constraints.

   iv.   Graph the inequalities and shade the solution set.

   v.   What does the shaded region represent?

   vi.   Suppose the manufacturer makes a profit of $10 on shirts and $18 on pajamas. How would it decide how many of each to make?

   vii.   How many of each should the manufacturer make, assuming it will sell all the shirts and pajamas it makes?

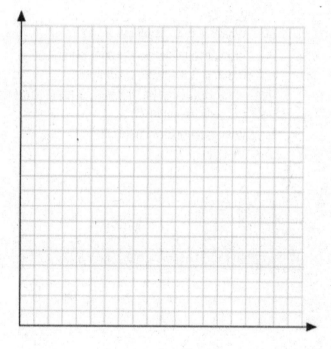

Problem Set

1. Find two numbers such that the sum of the first and three times the second is 5 and the sum of second and two times the first is 8.

2. A chemist has two solutions: a 50% methane solution and an 80% methane solution. He wants 100 mL of a 70% methane solution. How many mL of each solution does he need to mix?

3. Pam has two part time jobs. At one job, she works as a cashier and makes $8 per hour. At the second job, she works as a tutor and makes $12 per hour. One week she worked 30 hours and made $268. How many hours did she spend at each job?

4. A store sells Brazilian coffee for $10 per lb. and Columbian coffee for $14 per lb. If the store decides to make a 150-lb. blend of the two and sell it for $11 per lb., how much of each type of coffee should be used?

5. A potter is making cups and plates. It takes her 6 min. to make a cup and 3 min. to make a plate. Each cup uses $\frac{3}{4}$ lb. of clay, and each plate uses 1 lb. of clay. She has 20 hr. available to make the cups and plates and has 250 lb. of clay.
    a. What are the variables?
    b. Write inequalities for the constraints.
    c. Graph and shade the solution set.
    d. If she makes a profit of $2 on each cup and $1.50 on each plate, how many of each should she make in order to maximize her profit?
    e. What is her maximum profit?

Lesson 25: Solving Problems in Two Ways—Rates and Algebra

Exercise 1

a. Solve the following problem first using a tape diagram and then using an equation: In a school choir, $\frac{1}{2}$ of the members were girls. At the end of the year, 3 boys left the choir, and the ratio of boys to girls became $3\colon 4$. How many boys remained in the choir?

b. Which problem solution, the one using a tape diagram or the one using an equation, was easier to set up and solve? Why?

Mathematical Modeling Exercise/Exercise 2

Read the following problem:

All the printing presses at a print shop were scheduled to make copies of a novel and a cookbook. They were to print the same number of copies of each book, but the novel had twice as many pages as the cookbook. All of the printing presses worked for the first day on the larger book, turning out novels. Then, on day two, the presses were split into two equally sized groups. The first group continued printing copies of the novel and finished printing all the copies by the evening of the second day. The second group worked on the cookbook but did not finish by evening. One printing press, working for two additional full days, finished printing the remaining copies of the cookbooks. If all printing presses printed pages (for both the novel and cookbook) at the same constant rate, how many printing presses are there at the print shop?

    a.   Solve the problem working with rates to setup a tape diagram or an area model.

This work is derived from Eureka Math ™ and licensed by Great Minds. ©2015 Great Minds. eureka-math.org
ALG1-M1-SE-B1-1.3.0-05.2015

b. Solve the problem by setting up an equation.

Problem Set

1. Solve the following problems first using a tape diagram and then by setting up an equation. For each, give your opinion on which solution method was easier. Can you see the connection(s) between the two methods? What does each "unit" in the tape diagram stand for?

   a. 16 years from now, Pia's age will be twice her age 12 years ago. Find her present age.

   b. The total age of a woman and her son is 51 years. Three years ago, the woman was eight times as old as her son. How old is her son now?

   c. Five years from now, the sum of the ages of a woman and her daughter will be 40 years. The difference in their present age is 24 years. How old is her daughter now?

   d. Find three consecutive integers such that their sum is 51.

2. Solve the following problems by setting up an equation or inequality.

   a. If two numbers represented by $(2m + 1)$ and $(2m + 5)$ have a sum of 74, find m.

   b. Find two consecutive even numbers such that the sum of the smaller number and twice the greater number is 100.

   c. If 9 is subtracted from a number, and the result is multiplied by 19, the product is 171. Find the number.

   d. The product of two consecutive whole numbers is less than the sum of the square of the smaller number and 13.

3. The length, 18 meters, is the answer to the following question.

   "The length of a rectangle is three meters longer than its width. The area of the rectangle is 270 square meters. What is the length of the rectangle?"

   Rework this problem: Write an equation using L as the length (in meters) of the rectangle that would lead to the solution of the problem. Check that the answer above is correct by substituting 18 for L in your equation.

4. Jim tells you he paid a total of $23,078.90 for a car, and you would like to know the price of the car before sales tax so that you can compare the price of that model of car at various dealers. Find the price of the car before sales tax if Jim bought the car in each of the following states:

   a. Arizona, where the sales tax is 6.6%.

   b. New York, where the sales tax is 8.25%.

   c. A state where the sales tax is $s\%$.

5. A checking account is set up with an initial balance of $9,400, and $800 is removed from the account at the end of each month for rent. (No other user transactions occur on the account.)

   a. Write an inequality whose solutions are the months, m, in which the account balance is greater than $3,000. Write the solution set to your equation by identifying all of the solutions.

   b. Make a graph of the balance in the account after m months, and indicate on the plot the solutions to your inequality in part (a).

6. Axel and his brother like to play tennis. About three months ago they decided to keep track of how many games each has won. As of today, Axel has won 18 out of the 30 games against his brother.

   a. How many games would Axel have to win in a row in order to have a 75% winning record?

   b. How many games would Axel have to win in a row in order to have a 90% winning record?

   c. Is Axel ever able to reach a 100% winning record? Explain why or why not.

   d. Suppose that after reaching a winning record of 90% in part (b), Axel had a losing streak. How many games in a row would Axel have to lose in order to drop down to a winning record of 60% again?

7. Omar has $84 and Calina has $12. How much money must Omar give to Calina so that Calina will have three times as much as Omar?

   a. Solve the problem above by setting up an equation.

   b. In your opinion, is this problem easier to solve using an equation or using a tape diagram? Why?

Lesson 26: Recursive Challenge Problem—The Double and Add 5 Game

The *double and add 5* game is *loosely* related to the Collatz conjecture—an *unsolved* conjecture in mathematics named after Lothar Collatz, who first proposed the problem in 1937. The conjecture includes a recurrence relation, *triple and add 1,* as part of the problem statement. It is a worthwhile activity for you to read about the conjecture online.

Classwork

Example

Fill in the *doubling and adding* 5 below:

	Number	Double and add 5	
starting number →	1	$1 \cdot 2 + 5 = 7$	← result of round 1
	7	_____	← result of round 2
	____	_____	
	____	_____	
	____	_____	

Exercise 1

Complete the tables below for the given starting number.

Number	Double and add 5
2	_____
____	_____
____	_____

Number	Double and add 5
3	_____
____	_____
____	_____

Mathematical Modeling Exercise/Exercise 2

Given a starting number, double it and add 5 to get the result of round 1. Double the result of Round 1 and add 5, and so on. The goal of the game is to find the smallest starting whole number that produces a result of 100 or greater in three rounds or fewer.

Exercise 3

Using a generic initial value, a_0, and the recurrence relation, $a_{i+1} = 2a_i + 5$, for $i \geq 0$, find a formula for a_1, a_2, a_3, a_4 in terms of a_0.

Vocabulary

SEQUENCE: A *sequence* can be thought of as an ordered list of elements. The elements of the list are called the *terms of the sequence.*

For example, (P, O, O, L) is a sequence that is different than (L, O, O, P). Usually the terms are *indexed* (and therefore ordered) by a subscript starting at either 0 or 1: $a_1, a_2, a_3, a_4, \ldots$. The "…" symbol indicates that the pattern described is regular, that is, the next term is a_5, and the next is a_6, and so on. In the first example, $a_1 = P$ is the first term, $a_2 = O$ is the second term, and so on. Both finite and infinite sequences exist everywhere in mathematics. For example, the infinite decimal expansion of $\frac{1}{3} = 0.333333333\ldots$ can be represented as the sequence

(0.3, 0.33, 0.333, 0.3333, …).

RECURSIVE SEQUENCE: An example of a *recursive sequence* is a sequence that is defined by (1) specifying the values of one or more initial terms and (2) having the property that the remaining terms satisfy a recurrence relation that describes the value of a term based upon an algebraic expression in numbers, previous terms, or the index of the term.

The sequence generated by initial term, $a_1 = 3$, and recurrence relation, $a_n = 3a_{n-1}$, is the sequence (3, 9, 27, 81, 243, …). Another example, given by the initial terms, $a_0 = 1, a_1 = 1$, and recurrence relation, $a_n = a_{n-1} + a_{n-2}$, generates the famed *Fibonacci sequence* (1, 1, 2, 3, 5, …).

Problem Set

1. Write down the first 5 terms of the recursive sequences defined by the initial values and recurrence relations below:

   a. $a_0 = 0$ and $a_{i+1} = a_i + 1$, for $i \geq 0$,

   b. $a_1 = 1$ and $a_{i+1} = a_i + 1$, for $i \geq 1$,

   c. $a_1 = 2$ and $a_{i+1} = a_i + 2$, for $i \geq 1$,

   d. $a_1 = 3$ and $a_{i+1} = a_i + 3$, for $i \geq 1$,

   e. $a_1 = 2$ and $a_{i+1} = 2a_i$, for $i \geq 1$,

   f. $a_1 = 3$ and $a_{i+1} = 3a_i$, for $i \geq 1$,

   g. $a_1 = 4$ and $a_{i+1} = 4a_i$, for $i \geq 1$,

   h. $a_1 = 1$ and $a_{i+1} = (-1)a_i$, for $i \geq 1$,

   i. $a_1 = 64$ and $a_{i+1} = \left(-\frac{1}{2}\right)a_i$, for $i \geq 1$.

2. Look at the sequences you created in Problem 1 parts (b)–(d). How would you define a recursive sequence that generates multiples of 31?

3. Look at the sequences you created in Problem 1 parts (e)–(g). How would you define a recursive sequence that generates powers of 15?

4. The following recursive sequence was generated starting with an initial value of a_0 and the recurrence relation $a_{i+1} = 3a_i + 1$, for $i \geq 0$. Fill in the blanks of the sequence.

   (_____, _____, 94, _____, 850, _____).

5. For the recursive sequence generated by an initial value of a_0, and recurrence relation $a_{i+1} = a_i + 2$, for $i \geq 0$, find a formula for a_1, a_2, a_3, a_4 in terms of a_0. Describe in words what this sequence is generating.

6. For the recursive sequence generated by an initial value of a_0 and recurrence relation $a_{i+1} = 3a_i + 1$, for $i \geq 0$, find a formula for a_1, a_2, a_3, a_4 in terms of a_0.

This page intentionally left blank

Lesson 27: Recursive Challenge Problem—The Double and Add 5 Game

The *double and add 5* game is loosely related to the Collatz conjecture—an unsolved conjecture in mathematics named after Lothar Collatz, who first proposed the problem in 1937. The conjecture includes a recurrence relation, triple and add 1, as part of the problem statement. It is a worthwhile activity for you to read about the conjecture online.

Classwork

Example

Recall Exercise 3 from the previous lesson: Using a generic initial value, a_0, and the recurrence relation, $a_{i+1} = 2a_i + 5$, for $i \geq 0$, find a formula for a_1, a_2, a_3, a_4 in terms of a_0.

Mathematical Modeling Exercise/Exercise 1

Using one of the four formulas from Example 1, write an inequality that, if solved for a_0, will lead to finding the smallest starting whole number for the *double and add 5* game that produces a result of 1,000 or greater in 3 rounds or fewer.

Exercise 2

Solve the inequality derived in Exercise 1. Interpret your answer, and validate that it is the solution to the problem. That is, show that the whole number you found results in 1,000 or greater in three rounds, but the previous whole number takes four rounds to reach 1,000.

Exercise 3

Find the smallest starting whole number for the *double and add 5* game that produces a result of 1,000,000 or greater in four rounds or fewer.

Lesson Summary

The formula, $a_n = 2^n(a_0 + 5) - 5$, describes the nth term of the *double and add 5* game in terms of the starting number a_0 and n. Use this formula to find the smallest starting whole number for the *double and add 5* game that produces a result of 10,000,000 or greater in 15 rounds or fewer.

Problem Set

1. Your older sibling came home from college for the weekend and showed you the following sequences (from her homework) that she claimed were generated from initial values and recurrence relations. For each sequence, find an initial value and recurrence relation that describes the sequence. (Your sister showed you an answer to the first problem.)

   a. $(0, 2, 4, 6, 8, 10, 12, 14, 16, \ldots)$

   b. $(1, 3, 5, 7, 9, 11, 13, 15, 17, \ldots)$

   c. $(14, 16, 18, 20, 22, 24, 26, \ldots)$

   d. $(14, 21, 28, 35, 42, 49, \ldots)$

   e. $(14, 7, 0, -7, -14, -21, -28, -35, \ldots)$

   f. $(2, 4, 8, 16, 32, 64, 128, \ldots)$

   g. $(3, 6, 12, 24, 48, 96, \ldots)$

   h. $(1, 3, 9, 27, 81, 243, \ldots)$

   i. $(9, 27, 81, 243, \ldots)$

2. Answer the following questions about the recursive sequence generated by initial value, $a_1 = 4$, and recurrence relation, $a_{i+1} = 4a_i$ for $i \geq 1$.

   a. Find a formula for a_1, a_2, a_3, a_4, a_5 in terms of powers of 4.

   b. Your friend, Carl, says that he can describe the nth term of the sequence using the formula, $a_n = 4^n$. Is Carl correct? Write one or two sentences using the recurrence relation to explain why or why not.

3. The formula, $a_n = 2^n(a_0 + 5) - 5$, describes the nth term of the *double and add 5* game in terms of the starting number a_0 and n. Verify that it does describe the nth term by filling out the tables below for parts (b) through (e). (The first table is done for you.)

   a. Table for $a_0 = 1$

n	$2^n(a_0 + 5) - 5$
1	$2^1 \cdot 6 - 5 = 7$
2	$2^2 \cdot 6 - 5 = 19$
3	$2^3 \cdot 6 - 5 = 43$
4	$2^4 \cdot 6 - 5 = 91$

b. Table for $a_0 = 8$

n	$2^n(a_0 + 5) - 5$
1	_____
2	_____
3	_____
4	_____

c. Table for $a_0 = 9$

n	$2^n(a_0 + 5) - 5$
2	_____
3	_____

d. Table for $a_0 = 120$

n	$2^n(a_0 + 5) - 5$
3	_____
4	_____

e. Table for $a_0 = 121$

n	$2^n(a_0 + 5) - 5$
2	_____
3	_____

4. Bilbo Baggins stated to Samwise Gamgee, "Today, Sam, I will give you \$1. Every day thereafter for the next 14 days, I will take the previous day's amount, double it and add \$5, and give that new amount to you for that day."

   a. How much did Bilbo give Sam on day 15? (Hint: You don't have to compute each term.)

   b. Did Bilbo give Sam more than \$350,000 altogether?

5. The formula, $a_n = 2^{n-1}(a_0 + 5) - 5$, describes the nth term of the *double and add 5* game in terms of the starting number a_0 and n. Use this formula to find the smallest starting whole number for the *double and add 5* game that produces a result of 10,000,000 or greater in 15 rounds or fewer.

Lesson 27: Recursive Challenge Problem—The Double and Add 5 Game

EUREKA MATH™

Lesson 28: Federal Income Tax

Important Tax Tables for this Lesson

Exemption Deductions for Tax Year 2013

Exemption Class	Exemption Deduction
Single	$3,900
Married	$7,800
Married with 1 child	$11,700
Married with 2 children	$15,600
Married with 3 children	$19,500

Standard Deductions Based Upon Filing Status for Tax Year 2013

Filing Status	Standard Deduction
Single	$6,100
Married filing jointly	$12,200

Federal Income Tax for Married Filing Jointly for Tax Year 2013

If taxable income is over:	But not over--:	The tax is:	Plus the Marginal Rate	Of the amount over:
$0	$17,850	10%		$0
$17,850	$72,500	$1,785.00	15%	$17,850
$72,500	$146,400	$9,982.50	25%	$72,500
$146,400	$223,050	$28,457.50	28%	$146,400
$223,050	$398,350	$49,919.50	33%	$223,050
$398,350	$450,000	$107,768.50	35%	$398,350
$450,000 +		$125,846.00	39.6%	$450,000

TAXABLE INCOME: The U.S. government considers the *income* of a family (or individual) to include the sum of any money earned from a husband's or wife's jobs, and money made from their personal businesses or investments. The taxes for a household (i.e., an individual or family) are not computed from the income; rather, they are computed from the household's taxable income. For many families, the household's *taxable income* is simply the household's income minus exemption deductions and minus standard deductions:

(taxable income) = (income) − (exemption deduction) − (standard deduction)

All of the problems we will model in this lesson will use this equation to find a family's taxable income. The only exception is if the family's taxable income is less than zero, in which case we will say that the family's taxable income is just $0.

Use this formula and the tables above to answer the following questions about taxable income:

Exercise 1

Find the taxable income of a single person with no kids, who has an income of $55,000.

Exercise 2

Find the taxable income of a married couple with two children, who have a combined income of $55,000.

Exercise 3

Find the taxable income of a married couple with one child, who has a combined income of $23,000.

Lesson 28:   Federal Income Tax

Federal Income Tax and the Marginal Tax Rate: Below is an example of how to compute the federal income tax of a household using the Federal Income Tax table above.

Example 1

Compute the Federal Income Tax for the situation described in Exercise 1 (a single person with no kids making $55,000).

From the answer in Exercise 1, the taxable income is $45,000. Looking up $45,000 in the tax table above, we see that $45,000 corresponds to the second row because it is between $17,850 and $72,500:

If taxable income is over:	But not over:	The tax is:	Plus the Marginal Rate	Of the amount over:
$17,850	$72,500	$1,785.00	15%	$17,850

To calculate the tax, add $1,785 plus 15% of the amount of $45,000 that is over $17,850. Since $45000 - 17850 = 27150$, and 15% of 27,150 is $4,072.50, the total federal income tax on $45,000 of taxable income is $5,857.50.

Exercise 4

Compute the Federal Income Tax for a married couple with two children making $127,800.

Taxpayers sometimes misunderstand *marginal tax* to mean: "If my taxable income is $100,000, and my marginal tax rate is 25%, my federal income taxes are $25,000." This statement is not true—they would not owe $25,000 to the federal government. Instead, a marginal income tax charges a progressively higher tax rate for successively greater levels of income. Therefore, they would really owe:

- 10% on the first $17,850, or $1,785 in taxes for the interval from $0 to $17,850;
- 15% on the next $54,650, or $8,197.50 in taxes for the interval from $17,850 to $72,500;
- 25% on the last $27,500, or $6,875.00 in taxes for the interval from $72,500 to $100,000;

for a total of $16,857.50 of the $100,000 of taxable income. Thus, their *effective federal income tax rate* is 16.8575%, not 25% as they claimed. Note that the tax table above incorporates the different intervals so that only one calculation needs to be made (the answer to this problem is the same as the answer in Exercise 5).

Exercise 5

Create a table and a graph of federal income tax versus income for a married couple with two children between $0 of income and $500,000 of income.

Exercise 6

Interpret and validate the graph you created in Exercise 5. Does your graph provide an approximate value for the federal income tax you calculated in Exercise 4?

Exercise 7

Use the table you created in Exercise 5 to report on the effective federal income tax rate for a married couple with two children, who makes:

  a. $27,800

  b. $45,650

  c. $500,000

EUREKA
MATH™

Problem Set

Use the formula and tax tables provided in this lesson to perform all computations.

1. Find the taxable income of a married couple with two children, who have a combined income of $75,000.

2. Find the taxable income of a single person with no children, who has an income of $37,000.

3. Find the taxable income of a married couple with three children, who have a combined income of $62,000.

4. Find the federal income tax of a married couple with two children, who have a combined income of $100,000.

5. Find the federal income tax of a married couple with three children, who have a combined income of $300,000.

6. Find the effective federal income tax rate of a married couple with no children, who have a combined income of $34,000.

7. Find the effective federal income tax rate of a married couple with one child who have a combined income of $250,000.

8. The latest report on median household (family) income in the United States is $50,502 per year. Compute the federal income tax and effective federal income tax rate for a married couple with three children, who have a combined income of $50,502.

9. Extend the table you created in Exercise 6 by adding a column called, "Effective federal income tax rate." Compute the effective federal income tax rate to the nearest tenth for each row of the table, and create a graph that shows effective federal income tax rate versus income using the table.

This page intentionally left blank

Eureka Math
Algebra I
Module 2

Special thanks go to the Gordan A. Cain Center and to the Department of Mathematics at Louisiana State University for their support in the development of *Eureka Math*.

For a free *Eureka Math* Teacher
Resource Pack, Parent Tip
Sheets, and more please
visit www.Eureka.tools

Published by Great Minds

Copyright © 2015 Great Minds. All rights reserved. No part of this work may be reproduced or used in any form or by any means — graphic, electronic, or mechanical, including photocopying or information storage and retrieval systems — without written permission from the copyright holder. "Great Minds" and "Eureka Math" are registered trademarks of Great Minds.

Printed in the U.S.A.
This book may be purchased from the publisher at eureka-math.org
10 9 8 7 6 5 4 3 2 1

ISBN 978-1-63255-324-9

Lesson 1: Distributions and Their Shapes

Classwork

Statistics is all about data. Without data to talk about or to analyze or to question, statistics would not exist. There is a story to be uncovered behind all data—a story that has characters, plots, and problems. The questions or problems addressed by the data and their story can be disappointing, exciting, or just plain ordinary. This module is about stories that begin with data.

Example: Graphs

Data are often summarized by graphs; the graphs are the first indicator of variability in the data.

- **DOT PLOTS**: A plot of each data value on a scale or number line.

Dot Plot of Viewer Age

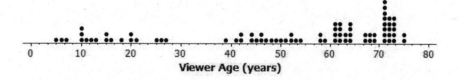

Viewer Age (years)

- **HISTOGRAMS**: A graph of data that groups the data based on intervals and represents the data in each interval by a bar.

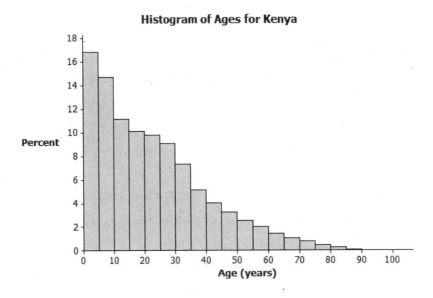

- **BOX PLOTS**: A graph that provides a picture of the data ordered and divided into four intervals that each contains approximately 25% of the data.

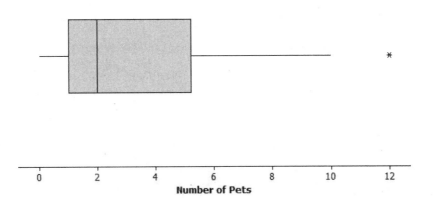

EUREKA
MATH™

Exercises

Answer the questions that accompany each graph to begin your understanding of the story behind the data.

Transportation officials collect data on flight delays (the number of minutes past the scheduled departure time that a flight takes off).

Consider the dot plot of the delay times for sixty BigAir flights during December 2012.

Dot Plot of December Delay Times

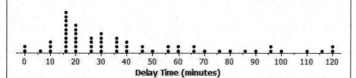

Delay Time (minutes)

1. What do you think this graph is telling us about the flight delays for these sixty flights?

2. Can you think of a reason why the data presented by this graph provide important information? Who might be interested in this data distribution?

3. Based on your previous work with dot plots, would you describe this dot plot as representing a symmetric or a skewed data distribution? (Recall that a skewed data distribution is not mound shaped.) Explain your answer.

A random sample of eighty viewers of a television show was selected. The dot plot below shows the distribution of the ages (in years) of these eighty viewers.

Dot Plot of Viewer Age

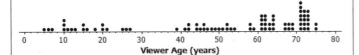

Viewer Age (years)

4. What do you think this graph is telling us about the ages of the eighty viewers in this sample?

5. Can you think of a reason why the data presented by this graph provide important information? Who might be interested in this data distribution?

6. Based on your previous work with dot plots, would you describe this dot plot as representing a symmetric or a skewed data distribution? Explain your answer.

The following histogram represents the age distribution of the population of Kenya in 2010.

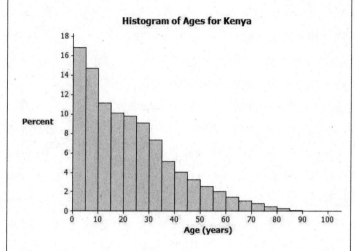

Histogram of Ages for Kenya

7. What do you think this graph is telling us about the population of Kenya?

8. Why might we want to study the data represented by this graph?

9. Based on your previous work with histograms, would you describe this histogram as representing a symmetrical or a skewed distribution? Explain your answer.

The following histogram represents the age distribution of the population of the United States in 2010.

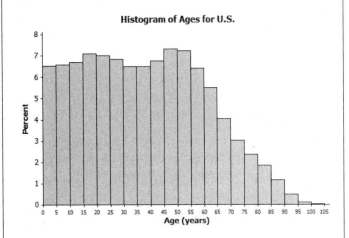

10. What do you think this graph is telling us about the population of the United States?

11. Why might we want to study the data represented by this graph?

Thirty students from River City High School were asked how many pets they owned. The following box plot was prepared from their answers.

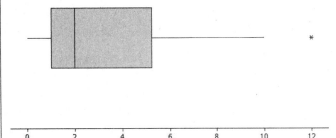

12. What does the box plot tell us about the number of pets owned by the thirty students at River City High School?

13. Why might understanding the data behind this graph be important?

Twenty-two juniors from River City High School participated in a walkathon to raise money for the school band. The following box plot was constructed using the number of miles walked by each of the twenty-two juniors.

Boxplot of Miles Walked for Juniors

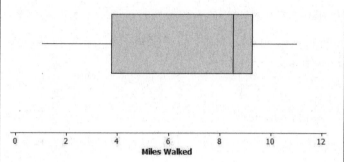

Miles Walked

14. What do you think the box plot tells us about the number of miles walked by the twenty-two juniors?

15. Why might understanding the data behind this graph be important?

Lesson Summary

Statistics is about data. Graphs provide a representation of the data distribution and are used to understand the data and to answer questions about the distribution.

Problem Set

1. Twenty-five people were attending an event. The ages of the people are as follows:

3, 3, 4, 4, 4, 4, 5, 6, 6, 6, 6, 6, 6, 6, 7, 7, 7, 7, 7, 7, 16, 17, 22, 22, 25.

   a. Create a histogram of the ages using the provided axes.

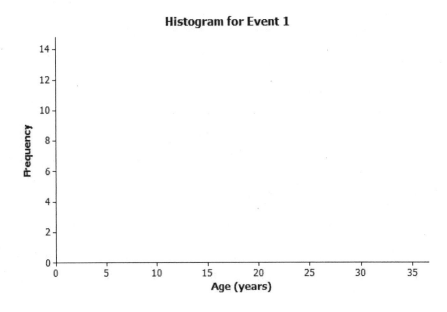

Histogram for Event 1

   b. Would you describe your graph as symmetrical or skewed? Explain your choice.

   c. Identify a typical age of the twenty-five people.

   d. What event do you think the twenty-five people were attending? Use your histogram to justify your conjecture.

EUREKA
MATH™

2. A different forty people were also attending an event. The ages of the people are as follows:

6, 13, 24, 27, 28, 32, 32, 34, 38, 42, 42, 43, 48, 49, 49, 49, 51, 52, 52, 53,

53, 53, 54, 55, 56, 57, 57, 60, 61, 61, 62, 66, 66, 66, 68, 70, 72, 78, 83, 97.

a. Create a histogram of the ages using the provided axes.

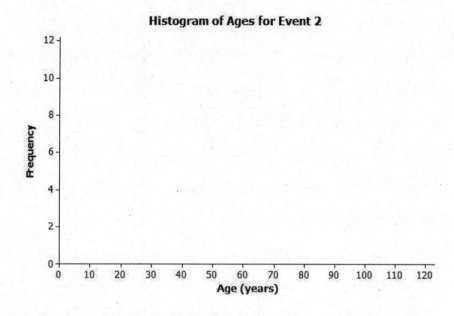

Histogram of Ages for Event 2

b. Would you describe your graph of ages as symmetrical or skewed? Explain your choice.

c. Identify a typical age of the forty people.

d. What event do you think the forty people were attending? Use your histogram to justify your conjecture.

e. How would you describe the differences in the two histograms?

EUREKA
MATH

Lesson 1:   Distributions and Their Shapes

S.9

This work is derived from Eureka Math ™ and licensed by Great Minds. ©2015 Great Minds. eureka-math.org
ALG1-M2-SE-B1-1.3.0-05.2015

This page intentionally left blank

Lesson 2: Describing the Center of a Distribution

Classwork

In previous work with data distributions, you learned how to derive the mean and the median of a data distribution. This lesson builds on your previous work with a center.

Exploratory Challenge/Exercises 1–9

Consider the following three sets of data.

Data Set 1: Pet owners

Students from River City High School were randomly selected and asked, "How many pets do you currently own?" The results are recorded below.

0	0	0	0	1	1	1	1	1	1	1	1	1	1	2
2	2	2	3	3	4	5	5	6	6	7	8	9	10	12

Data Set 2: Length of the east hallway at River City High School

Twenty students were selected to measure the length of the east hallway. Two marks were made on the hallway's floor: one at the front of the hallway and one at the end of the hallway. Each student was given a meter stick and asked to use the meter stick to determine the length between the marks to the nearest tenth of a meter. The results are recorded below.

8.2	8.3	8.3	8.4	8.4	8.5	8.5	8.5	8.5	8.5
8.6	8.6	8.6	8.6	8.7	8.7	8.8	8.8	8.9	8.9

Data Set 3: Age of cars

Twenty-five car owners were asked the age of their cars in years. The results are recorded below.

0	1	2	2	3	4	5	5	6	6	6	7	7
7	7	7	7	8	8	8	8	8	8	8	8	

1. Make a dot plot of each of the data sets. Use the following scales.

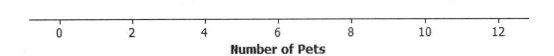

Number of Pets

East Hallway Length Measurement (meters)

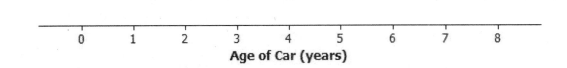

Age of Car (years)

2. Calculate the mean number of pets owned by the thirty students from River City High School. Calculate the median number of pets owned by the thirty students.

EUREKA
MATH™

3. What do you think is a typical number of pets for students from River City High School? Explain how you made your estimate.

4. Why do you think that different students got different results when they measured the same distance of the east hallway?

5. What is the mean length of the east hallway data set? What is the median length?

6. A construction company will be installing a handrail along a wall from the beginning point to the ending point of the east hallway. The company asks you how long the handrail should be. What would you tell the company? Explain your answer.

7. Describe the distribution of the age of cars.

8. What is the mean age of the twenty-five cars? What is the median age? Why are the mean and the median different?

9. What number would you use as an estimate of the typical age of a car for the twenty-five car owners? Explain your answer.

EUREKA MATH™

Lesson Summary

- A dot plot provides a graphical representation of a data distribution, helping us to visualize the distribution.

- The mean and the median of the distribution are numerical summaries of the center of a data distribution.

- When the distribution is nearly symmetrical, the mean and the median of the distribution are approximately equal. When the distribution is not symmetrical (often described as skewed), the mean and the median are not the same.

- For symmetrical distributions, the mean is an appropriate choice for describing a typical value for the distribution. For skewed data distributions, the median is a better description of a typical value.

Problem Set

Consider the following scenario. The company that created a popular video game "Leaders" plans to release a significant upgrade of the game. Users earn or lose points for making decisions as the leader of an imaginary country. In most cases, repeated playing of the game improves a user's ability to make decisions. The company will launch an online advertising campaign, but at the moment, they are not sure how to focus the advertising. Your goal is to help the company decide how the advertising campaign should be focused. Five videos have been proposed for the following target audiences:

    Video 1: Target females with beginning level scores

    Video 2: Target males with advanced level scores

    Video 3: Target all users with middle range level scores

    Video 4: Target males with beginning level scores

    Video 5: Target females with advanced level scores

1. Why might the company be interested in developing different videos based on user score?

2. Thirty female users and twenty-five male users were selected at random from a database of people who play the game regularly. Each of them agreed to be part of a research study and report their scores. A leadership score is based on a player's answers to leadership questions. A score of 1 to 40 is considered a beginning level leadership score, a score of 41 to 60 is considered a middle level leadership score, and a score of greater than 60 is considered an advanced level leadership score.

Use the following data to make a dot plot of the female scores, a dot plot of the male scores, and a dot plot of the scores for the combined group of males and females.

Female scores:

10	20	20	20	30	30	30	40	40	40
50	50	55	65	65	65	65	65	70	70
70	70	76	76	76	76	76	76	76	76

Male scores:

15	20	20	25	25	25	25	30	30	30
30	30	30	35	35	35	35	35	40	40
40	45	45	45	50					

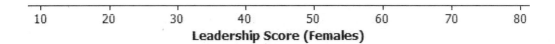

Leadership Score (Females)

Leadership Score (Males)

Leadership Score

EUREKA
MATH™

3. What do you think is a typical score for a female user? What do you think is a typical score for a male user? Explain how you determined these typical scores.

4. Why is it more difficult to report a typical score for the overall group that includes both the males and females?

5. Production costs will only allow for two video advertisements to be developed. Which two videos would you recommend for development? Explain your recommendations.

This page intentionally left blank

Lesson 3: Estimating Centers and Interpreting the Mean as a Balance Point

Classwork

Example

Your previous work in mathematics involved estimating a balance point of a data distribution. Let's review what we learned about the balance point of a distribution. A 12-inch ruler has several quarters taped to positions along the ruler. The broad side of a pencil is placed underneath the ruler to determine an approximate balance point of the ruler with the quarters.

Exercises 1–7

Consider the following example of quarters taped to a lightweight ruler.

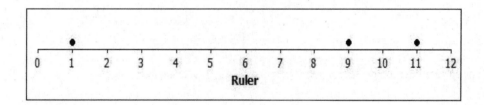

Ruler

1. Sam taped 3 quarters to his ruler. The quarters were taped to the positions 1 inch, 9 inches, and 11 inches. If the pencil was placed under the position 5 inches, do you think the ruler would balance? Why or why not?

2. If the ruler did not balance, would you move the pencil to the left or to the right of 5 inches to balance the ruler? Explain your answer.

3. Estimate a balance point for the ruler. Complete the following based on the position you selected.

Position of Quarter	Distance from Quarter to Your Estimate of the Balance Point
1	
9	
11	

4. What is the sum of the distances to the right of your estimate of the balance point?

5. What is the sum of the distances to the left of your estimate of the balance point?

6. Do you need to adjust the position of your balance point? If yes, explain how.

7. Calculate the mean and the median of the position of the quarters. Does the mean or the median of the positions provide a better estimate of the balance point for the position of the 3 quarters taped to this ruler? Explain why you made this selection.

EUREKA MATH™

Exercises 8–20

Twenty-two students from the junior class and twenty-six students from the senior class at River City High School participated in a walkathon to raise money for the school's band. Dot plots indicating the distances in miles students from each class walked are as follows.

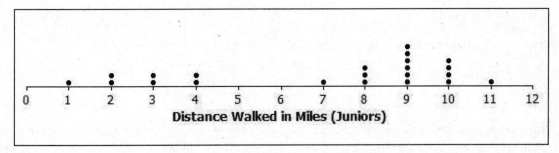

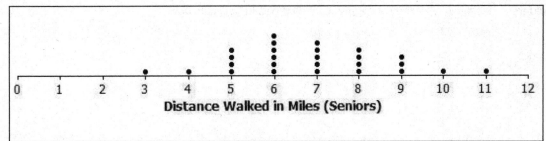

8. Estimate the mean number of miles walked by a junior, and mark it with an X on the junior class dot plot. How did you estimate this position?

9. What is the median of the junior data distribution?

10. Is the mean number of miles walked by a junior less than, approximately equal to, or greater than the median number of miles? If they are different, explain why. If they are approximately the same, explain why.

11. How would you describe the typical number of miles walked by a junior in this walkathon?

12. Estimate the mean number of miles walked by a senior, and mark it with an X on the senior class dot plot. How did you estimate this position?

13. What is the median of the senior data distribution?

14. Estimate the mean and the median of the miles walked by the seniors. Is your estimate of the mean number of miles less than, approximately equal to, or greater than the median number of miles walked by a senior? If they are different, explain why. If they are approximately the same, explain why.

15. How would you describe the typical number of miles walked by a senior in this walkathon?

16. A junior from River City High School indicated that the number of miles walked by a typical junior was better than the number of miles walked by a typical senior. Do you agree? Explain your answer.

Finally, the twenty-five sophomores who participated in the walkathon reported their results. A dot plot is shown below.

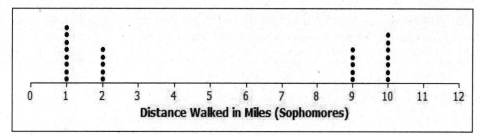

17. What is different about the sophomore data distribution compared to the data distributions for juniors and seniors?

18. Estimate the balance point of the sophomore data distribution.

19. What is the median number of miles walked by a sophomore?

20. How would you describe the sophomore data distribution?

> **Lesson Summary**
>
> The mean of a data distribution represents a balance point for the distribution. The sum of the distances to the right of the mean is equal to the sum of the distances to the left of the mean.

Problem Set

Consider another example of balance. Mr. Jackson is a mathematics teacher at Waldo High School. Students in his class are frequently given quizzes or exams. He indicated to his students that an exam is worth 4 quizzes when calculating an overall weighted average to determine their final grade. During one grading period, Scott got an 80% on one exam, a 90% on a second exam, a 60% on one quiz, and a 70% on another quiz.

How could we represent Scott's test scores? Consider the following number line.

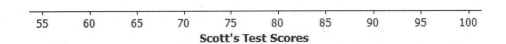

Scott's Test Scores

1. What values are represented by the number line?

2. If one "•" symbol is used to represent a quiz score, how might you represent an exam score?

3. Represent Scott's exams and quizzes on this number line using "•" symbols.

4. Mr. Jackson indicated that students should set an 85% overall weighted average as a goal. Do you think Scott met that goal? Explain your answer.

5. Place an X on the number line at a position that you think locates the balance point of all of the "•" symbols. Determine the sum of the distances from the X to each "•" on the right side of the X.

6. Determine the sum of the distances from the X to each "•" on the left side of the X.

7. Do the total distances to the right of the X equal the total distances to the left of the X?

8. Based on your answer to Problem 7, would you change your estimate of the balance point? If yes, where would you place your adjusted balance point? How does using this adjusted estimate change the total distances to the right of your estimate and the total distances to the left?

Lesson 3: Estimating Centers and Interpreting the Mean as a Balance Point

9. Scott's weighted average is 81. Recall that each exam score is equal to 4 times a quiz score. Show the calculations that lead to this weighted average.

10. How does the calculated mean score compare with your estimated balance point?

11. Compute the total distances to the right of the mean and the total distances to the left of the mean. What do you observe?

12. Did Scott achieve the goal set by Mr. Jackson of an 85% average? Explain your answer.

This page intentionally left blank

Lesson 4: Summarizing Deviations from the Mean

Classwork

Exercises 1–4

A consumers' organization is planning a study of the various brands of batteries that are available. As part of its planning, it measures lifetime (i.e., how long a battery can be used before it must be replaced) for each of six batteries of Brand A and eight batteries of Brand B. Dot plots showing the battery lives for each brand are shown below.

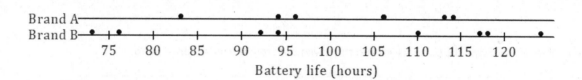

1. Does one brand of battery tend to last longer, or are they roughly the same? What calculations could you do in order to compare the battery lives of the two brands?

2. Do the battery lives tend to differ more from battery to battery for Brand A or for Brand B?

3. Would you prefer a battery brand that has battery lives that do not vary much from battery to battery? Why or why not?

The table below shows the lives (in hours) of the Brand A batteries.

Life (Hours)	83	94	96	106	113	114
Deviation from the Mean						

4. Calculate the deviations from the mean for the remaining values, and write your answers in the appropriate places in the table.

The table below shows the battery lives and the deviations from the mean for Brand B.

Life (Hours)	73	76	92	94	110	117	118	124
Deviation from the Mean	−27.5	−24.5	−8.5	−6.5	9.5	16.5	17.5	23.5

Exercises 5–10

The lives of five batteries of a third brand, Brand C, were determined. The dot plot below shows the lives of the Brand A and Brand C batteries.

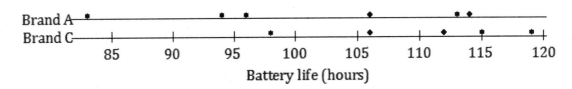

5. Which brand has the greater mean battery life? (You should be able to answer this question without doing any calculations.)

6. Which brand shows greater variability?

EUREKA
MATH™

7. Which brand would you expect to have the greater deviations from the mean (ignoring the signs of the deviations)?

The table below shows the lives for the Brand C batteries.

Life (Hours)	115	119	112	98	106
Deviation from the Mean					

8. Calculate the mean battery life for Brand C. (Be sure to include a unit in your answer.)

9. Write the deviations from the mean in the empty cells of the table for Brand C.

10. Ignoring the signs, are the deviations from the mean generally larger for Brand A or for Brand C? Does your answer agree with your answer to Exercise 7?

Exercises 11–15

The lives of 100 batteries of Brand D and 100 batteries of Brand E were determined. The results are summarized in the histograms below.

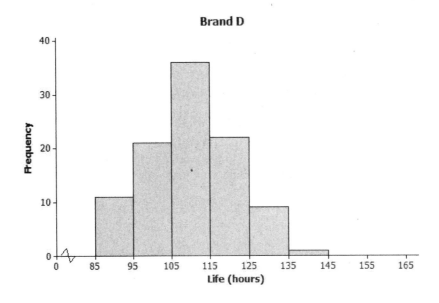

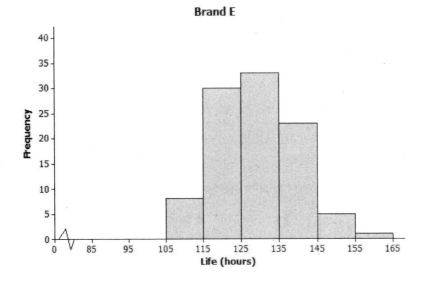

EUREKA
MATH™

11. Estimate the mean battery life for Brand D. (Do not do any calculations.)

12. Estimate the mean battery life for Brand E. (Do not do any calculations.)

13. Which of Brands D and E shows the greater variability in battery lives? Do you think the two brands are roughly the same in this regard?

14. Estimate the largest deviation from the mean for Brand D.

15. What would you consider a typical deviation from the mean for Brand D?

Lesson Summary

- For any given value in a data set, the deviation from the mean is the value minus the mean. Written algebraically, this is $x - \bar{x}$.
- The greater the variability (spread) of the distribution, the greater the deviations from the mean (ignoring the signs of the deviations).

Problem Set

1. Ten members of a high school girls' basketball team were asked how many hours they studied in a typical week. Their responses (in hours) were 20, 13, 10, 6, 13, 10, 13, 11, 11, 10.

   a. Using the axis given below, draw a dot plot of these values. (Remember, when there are repeated values, stack the dots with one above the other.)

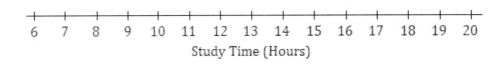

Study Time (Hours)

   b. Calculate the mean study time for these students.

   c. Calculate the deviations from the mean for these study times, and write your answers in the appropriate places in the table below.

Number of Hours Studied	20	13	10	6	13	10	13	11	11	10
Deviation from the Mean										

   d. The study times for fourteen girls from the soccer team at the same school as the one above are shown in the dot plot below.

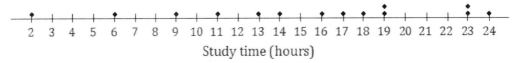

Study time (hours)

   Based on the data, would the deviations from the mean (ignoring the sign of the deviations) be greater or less for the soccer players than for the basketball players?

EUREKA MATH™

2. All the members of a high school softball team were asked how many hours they studied in a typical week. The results are shown in the histogram below.

(The data set in this question comes from NCTM Core Math Tools,

http://www.nctm.org/Classroom-Resources/Core-Math-Tools/Data-Sets/)

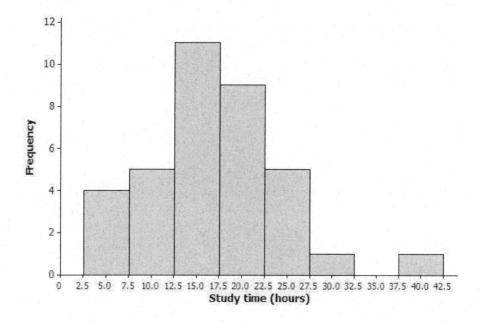

a. We can see from the histogram that four students studied around 5 hours per week. How many students studied around 15 hours per week?

b. How many students were there in total?

c. Suppose that the four students represented by the histogram bar centered at 5 had all studied exactly 5 hours, the five students represented by the next histogram bar had all studied exactly 10 hours, and so on. If you were to add up the study times for all of the students, what result would you get?

d. What is the mean study time for these students?

e. What would you consider to be a typical deviation from the mean for this data set?

This page intentionally left blank

Lesson 5: Measuring Variability for Symmetrical Distributions

Classwork

Example 1: Calculating the Standard Deviation

Here is a dot plot of the lives of the Brand A batteries from Lesson 4.

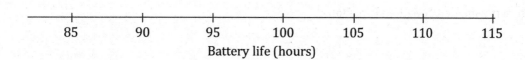

Battery life (hours)

How do you measure variability of this data set? One way is by calculating **standard deviation**.

- First, find each deviation from the mean.

- Then, square the deviations from the mean. For example, when the deviation from the mean is -18, the squared deviation from the mean is $(-18)^2 = 324$.

Life (Hours)	83	94	96	106	113	114
Deviation from the Mean	-18	-7	-5	5	12	13
Squared Deviations from the Mean	324	49	25	25	144	169

- Add up the squared deviations:

    $$324 + 49 + 25 + 25 + 144 + 169 = 736.$$

    This result is the *sum* of the squared deviations.

The number of values in the data set is denoted by n. In this example, n is 6.

- You divide the sum of the squared deviations by $n - 1$, which here is $6 - 1$, or 5.

    $$\frac{736}{5} = 147.2$$

- Finally, you take the square root of 147.2, which to the nearest hundredth is 12.13.

That is the standard deviation! It seems like a very complicated process at first, but you will soon get used to it.

We conclude that a typical deviation of a Brand A battery lifetime from the mean battery lifetime for Brand A is 12.13 hours. The unit of standard deviation is always the same as the unit of the original data set. So, the standard deviation to the nearest hundredth, with the unit, is 12.13 hours. How close is the answer to the typical deviation that you estimated at the beginning of the lesson?

Exercises 1–5

Now you can calculate the standard deviation of the lifetimes for the eight Brand B batteries. The mean was 100.5. We already have the deviations from the mean.

Life (Hours)	73	76	92	94	110	117	118	124
Deviation from the Mean	-27.5	-24.5	-8.5	-6.5	9.5	16.5	17.5	23.5
Squared Deviation from the Mean								

1. Write the squared deviations in the table.

2. Add up the squared deviations. What result do you get?

3. What is the value of n for this data set? Divide the sum of the squared deviations by $n - 1$, and write your answer below. Round your answer to the nearest thousandth.

4. Take the square root to find the standard deviation. Record your answer to the nearest hundredth.

5. How would you interpret the standard deviation that you found in Exercise 4? (Remember to give your answer in the context of this question. Interpret your answer to the nearest hundredth.)

Exercises 6–7

Jenna has bought a new hybrid car. Each week for a period of seven weeks, she has noted the fuel efficiency (in miles per gallon) of her car. The results are shown below.

$$45 \quad 44 \quad 43 \quad 44 \quad 45 \quad 44 \quad 43$$

6. Calculate the standard deviation of these results to the nearest hundredth. Be sure to show your work.

7. What is the meaning of the standard deviation you found in Exercise 6?

Lesson Summary

- The standard deviation measures a typical deviation from the mean.
- To calculate the standard deviation,
    1. Find the mean of the data set;
    2. Calculate the deviations from the mean;
    3. Square the deviations from the mean;
    4. Add up the squared deviations;
    5. Divide by $n - 1$ (if working with data from a sample, which is the most common case);
    6. Take the square root.
- The unit of the standard deviation is always the same as the unit of the original data set.
- The larger the standard deviation, the greater the spread (variability) of the data set.

Problem Set

1. A small car dealership tests the fuel efficiency of sedans on its lot. It chooses 12 sedans for the test. The fuel efficiency (mpg) values of the cars are given in the table below. Complete the table as directed below.

Fuel Efficiency (miles per gallon)	29	35	24	25	21	21	18	28	31	26	26	22
Deviation from the Mean												
Squared Deviation from the Mean												

   a. Calculate the mean fuel efficiency for these cars.
   b. Calculate the deviations from the mean, and write your answers in the second row of the table.
   c. Square the deviations from the mean, and write the squared deviations in the third row of the table.
   d. Find the sum of the squared deviations.
   e. What is the value of n for this data set? Divide the sum of the squared deviations by $n - 1$.
   f. Take the square root of your answer to part (e) to find the standard deviation of the fuel efficiencies of these cars. Round your answer to the nearest hundredth.

Lesson 5:   Measuring Variability for Symmetrical Distributions

2. The same dealership decides to test fuel efficiency of SUVs. It selects six SUVs on its lot for the test. The fuel efficiencies (in miles per gallon) of these cars are shown below.

21   21   21   30   28   24

Calculate the mean and the standard deviation of these values. Be sure to show your work, and include a unit in your answer.

3. Consider the following questions regarding the cars described in Problems 1 and 2.

   a. What is the standard deviation of the fuel efficiencies of the cars in Problem 1? Explain what this value tells you.

   b. You also calculated the standard deviation of the fuel efficiencies for the cars in Problem 2. Which of the two data sets (Problem 1 or Problem 2) has the larger standard deviation? What does this tell you about the two types of cars (sedans and SUVs)?

This page intentionally left blank

Lesson 6: Interpreting the Standard Deviation

Classwork

Example 1

Your teacher will show you how to use a calculator to find the mean and standard deviation for the following set of data.

A set of eight men have heights (in inches) as shown below.

67.0   70.9   67.6   69.8   69.7   70.9   68.7   67.2

Indicate the mean and standard deviation you obtained from your calculator to the nearest hundredth.

Mean: _____

Standard Deviation: _____

Exercise 1

1. The heights (in inches) of nine women are as shown below.

68.4   70.9   67.4   67.7   67.1   69.2   66.0   70.3   67.6

Use the statistical features of your calculator or computer software to find the mean and the standard deviation of these heights to the nearest hundredth.

Mean: _____

Standard Deviation: _____

Exploratory Challenge/Exercises 2–5

2. A group of people attended a talk at a conference. At the end of the talk, ten of the attendees were given a questionnaire that consisted of four questions. The questions were optional, so it was possible that some attendees might answer none of the questions, while others might answer 1, 2, 3, or all 4 of the questions (so, the possible numbers of questions answered are 0, 1, 2, 3, and 4).

    Suppose that the numbers of questions answered by each of the ten people were as shown in the dot plot below.

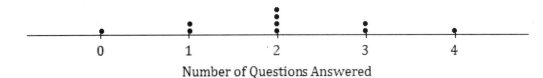

Number of Questions Answered

Use the statistical features of your calculator to find the mean and the standard deviation of the data set.

Mean: _____

Standard Deviation: _____

3. Suppose the dot plot looked like this:

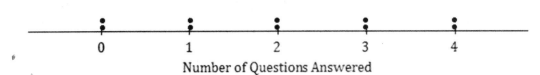

Number of Questions Answered

a. Use your calculator to find the mean and the standard deviation of this distribution.

b. Remember that the size of the standard deviation is related to the size of the deviations from the mean. Explain why the standard deviation of this distribution is greater than the standard deviation in Exercise 2.

EUREKA
MATH™

4. Suppose that all ten people questioned answered all four questions on the questionnaire.

   a. What would the dot plot look like?

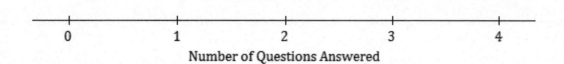

   b. What is the mean number of questions answered? (You should be able to answer without doing any calculations!)

   c. What is the standard deviation? (Again, don't do any calculations!)

5. Continue to think about the situation previously described where the numbers of questions answered by each of ten people was recorded.

   a. Draw the dot plot of the distribution of possible data values that has the largest possible standard deviation. (There were ten people at the talk, so there should be ten dots in your dot plot.) Use the scale given below.

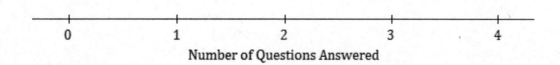

   b. Explain why the distribution you have drawn has a larger standard deviation than the distribution in Exercise 4.

Lesson Summary

- The mean and the standard deviation of a data set can be found directly using the statistical features of a calculator.
- The size of the standard deviation is related to the sizes of the deviations from the mean. Therefore, the standard deviation is minimized when all the numbers in the data set are the same and is maximized when the deviations from the mean are made as large as possible.

Problem Set

1. At a track meet, there are three men's 100 m races. The times for eight of the sprinters are recorded to the nearest $\frac{1}{10}$ of a second. The results of the three races for these eight sprinters are shown in the dot plots below.

Race 1

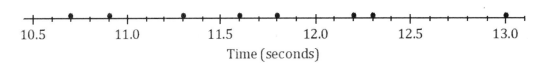

Race 2

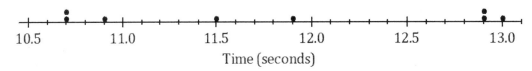

Race 3

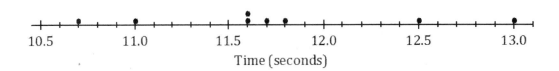

a. Remember that the size of the standard deviation is related to the sizes of the deviations from the mean. Without doing any calculations, indicate which of the three races has the smallest standard deviation of times. Justify your answer.

b. Which race had the largest standard deviation of times? (Again, don't do any calculations!) Justify your answer.

c. Roughly what would be the standard deviation in Race 1? (Remember that the standard deviation is a typical deviation from the mean. So, here you are looking for a typical deviation from the mean, in seconds, for Race 1.)

Lesson 6:   Interpreting the Standard Deviation

d. Use your calculator to find the mean and the standard deviation for each of the three races. Write your answers in the table below to the nearest thousandth.

	Mean	**Standard Deviation**
Race 1		
Race 2		
Race 3		

e. How close were your answers for parts (a)–(c) to the actual values?

2. A large city, which we will call City A, holds a marathon. Suppose that the ages of the participants in the marathon that took place in City A were summarized in the histogram below.

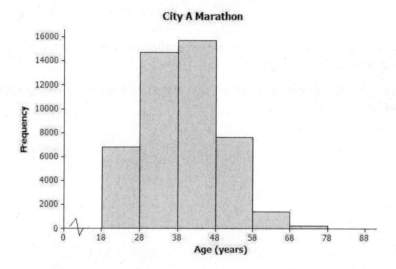

a. Make an estimate of the mean age of the participants in the City A marathon.

b. Make an *estimate* of the standard deviation of the ages of the participants in the City A marathon.

A smaller city, City B, also held a marathon. However, City B restricts the number of people of each age category who can take part to 100. The ages of the participants are summarized in the histogram below.

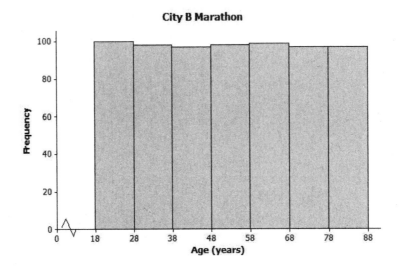

c. Approximately what was the mean age of the participants in the City B marathon? Approximately what was the standard deviation of the ages?

d. Explain why the standard deviation of the ages in the City B marathon is greater than the standard deviation of the ages for the City A marathon.

Lesson 7: Measuring Variability for Skewed Distributions (Interquartile Range)

Classwork

Exploratory Challenge 1/Exercises 1–3: Skewed Data and Their Measure of Center

Consider the following scenario. A television game show, *Fact or Fiction*, was cancelled after nine shows. Many people watched the nine shows and were rather upset when it was taken off the air. A random sample of eighty viewers of the show was selected. Viewers in the sample responded to several questions. The dot plot below shows the distribution of ages of these eighty viewers.

Dot Plot of Viewer Age

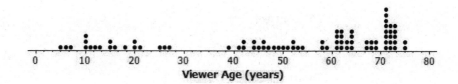

1. Approximately where would you locate the mean (balance point) in the above distribution?

2. How does the direction of the tail affect the location of the mean age compared to the median age?

3. The mean age of the above sample is approximately 50. Do you think this age describes the typical viewer of this show? Explain your answer.

Exploratory Challenge 2/Exercises 4–8: Constructing and Interpreting the Box Plot

4. Using the above dot plot, construct a box plot over the dot plot by completing the following steps:
    i.   Locate the middle 40 observations, and draw a box around these values.
    ii. Calculate the median, and then draw a line in the box at the location of the median.
    iii. Draw a line that extends from the upper end of the box to the largest observation in the data set.
    iv. Draw a line that extends from the lower edge of the box to the minimum value in the data set.

5. Recall that the 5 values used to construct the dot plot make up the 5-number summary. What is the 5-number summary for this data set of ages?

    Minimum age: _____

    Lower quartile or Q1:   _____

    Median age: _____

    Upper quartile or Q3:   _____

    Maximum age: _____

6. What percent of the data does the box part of the box plot capture?

7. What percent of the data fall between the minimum value and Q1?

8. What percent of the data fall between Q3 and the maximum value?

Exercises 9–14

An advertising agency researched the ages of viewers most interested in various types of television ads. Consider the following summaries:

Ages	Target Products or Services
30–45	Electronics, home goods, cars
46–55	Financial services, appliances, furniture
56–72	Retirement planning, cruises, health-care services

9. The mean age of the people surveyed is approximately 50 years old. As a result, the producers of the show decided to obtain advertisers for a typical viewer of 50 years old. According to the table, what products or services do you think the producers will target? Based on the sample, what percent of the people surveyed about the *Fact or Fiction* show would have been interested in these commercials if the advertising table is accurate?

10. The show failed to generate the interest the advertisers hoped. As a result, they stopped advertising on the show, and the show was cancelled. Kristin made the argument that a better age to describe the typical viewer is the median age. What is the median age of the sample? What products or services does the advertising table suggest for viewers if the median age is considered as a description of the typical viewer?

11. What percent of the people surveyed would be interested in the products or services suggested by the advertising table if the median age were used to describe a typical viewer?

12. What percent of the viewers have ages between Q1 and Q3? The difference between Q3 and Q1, or Q3 − Q1, is called the interquartile range, or IQR. What is the IQR for this data distribution?

13. The IQR provides a summary of the variability for a skewed data distribution. The IQR is a number that specifies the length of the interval that contains the middle half of the ages of viewers. Do you think producers of the show would prefer a show that has a small or large interquartile range? Explain your answer.

14. Do you agree with Kristin's argument that the median age provides a better description of a typical viewer? Explain your answer.

Exploratory Challenge 3/Exercises 15–20: Outliers

Students at Waldo High School are involved in a special project that involves communicating with people in Kenya. Consider a box plot of the ages of 200 randomly selected people from Kenya.

Box Plot of Ages for Kenya

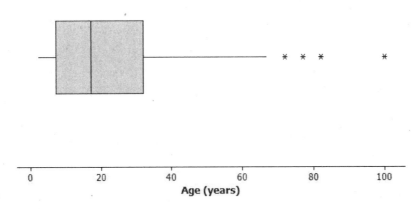

A data distribution may contain extreme data (specific data values that are unusually large or unusually small relative to the median and the interquartile range). A box plot can be used to display extreme data values that are identified as outliers.

Each "*" in the box plot represents the ages of four people from this sample. Based on the sample, these four ages were considered outliers.

EUREKA MATH™

15. Estimate the values of the four ages represented by an *.

An outlier is defined to be any data value that is more than $1.5 \times (IQR)$ away from the nearest quartile.

16. What is the median age of the sample of ages from Kenya? What are the approximate values of Q1 and Q3? What is the approximate IQR of this sample?

17. Multiply the IQR by 1.5. What value do you get?

18. Add $1.5 \times (IQR)$ to the third quartile age (Q3). What do you notice about the four ages identified by an *?

19. Are there any age values that are less than $Q1 - 1.5 \times (IQR)$? If so, these ages would also be considered outliers.

20. Explain why there is no * on the low side of the box plot for ages of the people in the sample from Kenya.

Lesson Summary

- Nonsymmetrical data distributions are referred to as skewed.
- Left-skewed or skewed to the left means the data spread out longer (like a tail) on the left side.
- Right-skewed or skewed to the right means the data spread out longer (like a tail) on the right side.
- The center of a skewed data distribution is described by the median.
- Variability of a skewed data distribution is described by the interquartile range (IQR).
- The IQR describes variability by specifying the length of the interval that contains the middle 50% of the data values.
- Outliers in a data set are defined as those values more than $1.5 \times (IQR)$ from the nearest quartile. Outliers are usually identified by an "*" or a "•" in a box plot.

Problem Set

Consider the following scenario. Transportation officials collect data on flight delays (the number of minutes a flight takes off after its scheduled time).

Consider the dot plot of the delay times in minutes for 60 BigAir flights during December 2012:

Dot Plot of December Delay Times

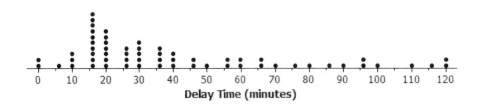

1. How many flights left more than 60 minutes late?

2. Why is this data distribution considered skewed?

3. Is the tail of this data distribution to the right or to the left? How would you describe several of the delay times in the tail?

4. Draw a box plot over the dot plot of the flights for December.

5. What is the interquartile range, or IQR, of this data set?

6. The mean of the 60 flight delays is approximately 42 minutes. Do you think that 42 minutes is typical of the number of minutes a BigAir flight was delayed? Why or why not?

7. Based on the December data, write a brief description of the BigAir flight distribution for December.

8. Calculate the percentage of flights with delays of more than 1 hour. Were there many flight delays of more than 1 hour?

9. BigAir later indicated that there was a flight delay that was not included in the data. The flight not reported was delayed for 48 hours. If you had included that flight delay in the box plot, how would you have represented it? Explain your answer.

10. Consider a dot plot and the box plot of the delay times in minutes for 60 BigAir flights during January 2013.

    How is the January flight delay distribution different from the one summarizing the December flight delays? In terms of flight delays in January, did BigAir improve, stay the same, or do worse compared to December? Explain your answer.

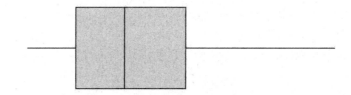

Delay Time (January)

Box Plot of January Delay Times

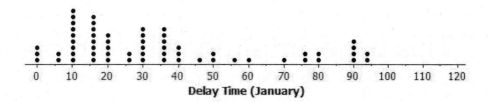

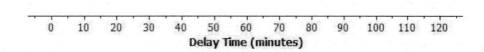

Delay Time (minutes)

This page intentionally left blank

Lesson 8: Comparing Distributions

Classwork

Exploratory Challenge 1: Country Data

A science museum has a Traveling Around the World exhibit. Using 3D technology, participants can make a virtual tour of cities and towns around the world. Students at Waldo High School registered with the museum to participate in a virtual tour of Kenya, visiting the capital city of Nairobi and several small towns. Before they take the tour, however, their mathematics class decided to study Kenya using demographic data from 2010 provided by the United States Census Bureau. They also obtained data for the United States from 2010 to compare to data for Kenya.

The following histograms represent the age distributions of the two countries.

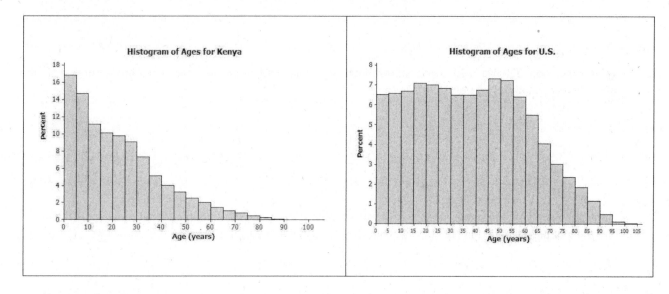

Exercises 1–8

1. How do the shapes of the two histograms differ?

2. Approximately what percent of people in Kenya are between the ages of 0 and 10 years?

3. Approximately what percent of people in the United States are between the ages of 0 and 10 years?

4. Approximately what percent of people in Kenya are 60 years or older?

5. Approximately what percent of people in the United States are 60 years or older?

6. The population of Kenya in 2010 was approximately 41 million people. What is the approximate number of people in Kenya between the ages of 0 and 10 years?

7. The population of the United States in 2010 was approximately 309 million people. What is the approximate number of people in the United States between the ages of 0 and 10 years?

8. The Waldo High School students started planning for their virtual visit of the neighborhoods in Nairobi and several towns in Kenya. Do you think they will see many teenagers? Will they see many senior citizens who are 70 or older? Explain your answer based on the histogram.

Lesson 8:   Comparing Distributions

EUREKA
MATH™

Exploratory Challenge 2: Learning More About the Countries Using Box Plots and Histograms

A random sample of 200 people from Kenya in 2010 was discussed in previous lessons. A random sample of 200 people from the United States is also available for study. Box plots constructed using the ages of the people in these two samples are shown below.

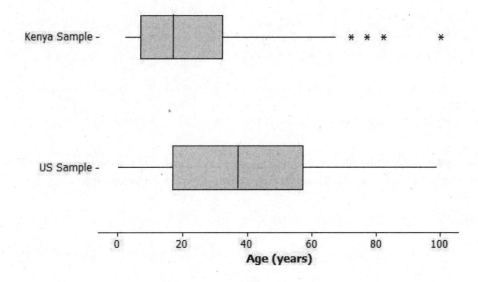

Exercises 9–16

9. Adrian, a senior at Waldo High School, stated that the box plots indicate that the United States has a lot of older people compared to Kenya. Would you agree? How would you describe the difference in the ages of people in these two countries based on the above box plots?

10. Estimate the median age of a person in Kenya and the median age of a person in the United States using the box plots.

11. Using the box plot, 25% of the people in the United States are younger than what age? How did you determine that age?

12. Using the box plot, approximately what percent of people in Kenya are younger than 18 years old?

13. Could you have estimated the mean age of a person from Kenya using the box plot? Explain your answer.

14. The mean age of people the United States is approximately 38 years. Using the histogram, estimate the percent of people in the United States who are younger than the mean age in the United States.

15. If the median age is used to describe a typical person in Kenya, what percent of people in Kenya are younger than the median age? Is the mean or median age a better description of a typical person in Kenya? Explain your answer.

16. What is the IQR of the ages in the sample from the United States? What is the IQR of the ages in the sample from Kenya? If the IQRs are used to compare countries, what does a smaller IQR indicate about a country? Use Kenya and the United States to explain your answer.

Lesson Summary

- Histograms show the general shape of a distribution.

- Box plots are created from the 5-number summary of a data set.

- A box plot identifies the median, minimum, and maximum values and the upper and lower quartiles.

- The interquartile range (IQR) describes how the data are spread around the median; it is the length of the interval that contains 50% of the data values.

- The median is used as a measure of the center when a distribution is skewed or contains outliers.

Problem Set

The following box plot summarizes ages for a random sample from a made-up country named Math Country.

Boxplot of Ages for Sample From Math Country

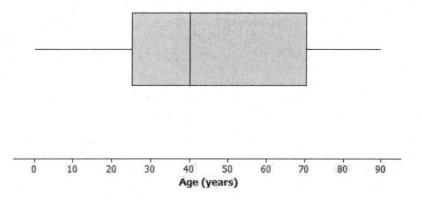

1. Make up your own sample of forty ages that could be represented by the box plot for Math Country. Use a dot plot to represent the ages of the forty people in Math Country.

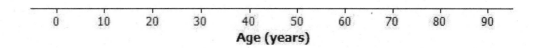

2. Is the sample of forty ages represented in your dot plot of Math Country the only sample that could be represented by the box plot? Explain your answer.

3. The following is a dot plot of sixty ages from a random sample of people from Japan in 2010. Draw a box plot over this dot plot.

4. Based on your box plot, would the median age of people in Japan be closer to the median age of people in Kenya or the United States? Justify your answer.

5. What does the box plot of this sample from Japan indicate about the possible differences in the age distributions of people from Japan and Kenya?

Lesson 9: Summarizing Bivariate Categorical Data

Classwork

Recall from your work in Grade 6 and Grade 8 that categorical data are data that are not numbers. Bivariate categorical data results from collecting data on two categorical variables. In this lesson, you will see examples involving categorical data collected from two survey questions.

Exploratory Challenge 1: Superhero Powers

Superheroes have been popular characters in movies, television, books, and comics for many generations. Superman was one of the most popular series in the 1950s while Batman was a top-rated series in the 1960s. Each of these characters was also popular in movies released from 1990 to 2013. Other notable characters portrayed in movies over the last several decades include Captain America, She-Ra, and the Fantastic Four. What is special about a superhero? Is there a special superhero power that makes these characters particularly popular?

High school students in the United States were invited to complete an online survey in 2010. Part of the survey included questions about superhero powers. More than 1,000 students responded to this survey that included a question about a favorite superhero power. Researchers randomly selected 450 of the completed surveys. A rather confusing breakdown of the data by gender was compiled from the 450 surveys:

- 100 students indicated their favorite power was to fly. 49 of those students were females.
- 131 students selected the power to freeze time as their favorite power. 71 of those students were males.
- 75 students selected invisibility as their favorite power. 48 of those students were females.
- 26 students indicated super strength as their favorite power. 25 of those students were males.
- And finally, 118 students indicated telepathy as their favorite power. 70 of those students were females.

Exercises 1–4

Several superheroes portrayed in movies and television series had at least one extraordinary power. Some superheroes had more than one special power. Was Superman's power to fly the favorite power of his fans, or was it his super strength? Would females view the power to fly differently than males, or in the same way? Use the survey information given in Example 1 to answer the following questions.

1. How many more females than males indicated their favorite power is telepathy?

2. How many more males than females indicated their favorite power was to fly?

3. Write survey questions that you think might have been used to collect this data.

4. How do you think the 450 surveys used in Example 1 might have been selected? You can assume that there were 1,000 surveys to select from.

Exploratory Challenge 2: A Statistical Study Involving a Two-Way Frequency Table

The data in Example 1 prompted students in a mathematics class to pose the statistical question, "Do high school males have different preferences for superhero powers than high school females?" Answering this statistical question involves collecting data as well as anticipating variability in the data collected.

The data consist of two responses from each student completing a survey. The first response indicates a student's gender, and the second response indicates the student's favorite superpower. For example, data collected from one student was *male* and *to fly*. The data are bivariate categorical data.

The first step in analyzing the statistical question posed by the students in their mathematics class is to organize this data in a two-way frequency table.

A two-way frequency table that can be used to organize the categorical data is shown below. The letters below represent the frequency counts of the cells of the table.

	To Fly	Freeze time	Invisibility	Super Strength	Telepathy	Total
Females	(a)	(b)	(c)	(d)	(e)	(f)
Males	(g)	(h)	(i)	(j)	(k)	(l)
Total	(m)	(n)	(o)	(p)	(q)	(r)

- The shaded cells are called *marginal frequencies*. They are located around the margins of the table and represent the totals of the rows or columns of the table.

- The non-shaded cells *within* the table are called *joint frequencies*. Each joint cell is the frequency count of responses from the two categorical variables located by the intersection of a row and column.

Exercises 5–12

5. Describe the data that would be counted in cell (a).

6. Describe the data that would be counted in cell (j).

7. Describe the data that would be counted in cell (l).

8. Describe the data that would be counted in cell (n).

9. Describe the data that would be counted in cell (r).

10. Cell (i) is the number of male students who selected *invisibility* as their favorite superpower. Using the information given in Example 1, what is the value of this number?

11. Cell (d) is the number of females whose favorite superpower is super strength. Using the information given in Example 1, what is the value of this number?

EUREKA
MATH™

Lesson 9:   Summarizing Bivariate Categorical Data

S.63

This work is derived from Eureka Math ™ and licensed by Great Minds. ©2015 Great Minds. eureka-math.org
ALG1-M2-SE-B1-1.3.0-05.2015

12. Complete the table below by determining a frequency count for each cell based on the summarized data.

	To Fly	Freeze Time	Invisibility	Super Strength	Telepathy	Total
Females						
Males						
Total						

Lesson Summary

- *Categorical data* are data that take on values that are categories rather than numbers. Examples include male or female for the categorical variable of gender or the five superpower categories for the categorical variable of superpower qualities.
- A *two-way frequency table* is used to summarize bivariate categorical data.
- The number in a two-way frequency table at the intersection of a row and column of the response to two categorical variables represents a *joint frequency*.
- The total number of responses for each value of a categorical variable in the table represents the *marginal frequency* for that value.

Problem Set

Several students at Rufus King High School were debating whether males or females were more involved in after-school activities. There are three organized activities in the after-school program—intramural basketball, chess club, and jazz band. Due to budget constraints, a student can only select one of these activities. The students were not able to ask every student in the school whether they participated in the after-school program or what activity they selected if they were involved.

1. Write questions that could be included in the survey to investigate the question the students are debating. Questions that could be used for this study include the following:

2. Rufus King High School has approximately 1,500 students. Sam suggested that the first 100 students entering the cafeteria for lunch would provide a random sample to analyze. Janet suggested that they pick 100 students based on a school identification number. Who has a better strategy for selecting a random sample? How do you think 100 students could be randomly selected to complete the survey?

3. Consider the following results from 100 randomly selected students:
   - Of the 60 female students selected, 20 of them played intramural basketball, 10 played chess, and 10 were in the jazz bland. The rest of them did not participate in the after-school program.
   - Of the male students, 10 did not participate in the after-school program, 20 played intramural basketball, 8 played in the jazz band, and the rest played chess.

   A two-way frequency table to summarize the survey data was started. Indicate what label is needed in the table cell identified with a ???.

	Intramural Basketball	Chess Club	Jazz Band	???	Total
Female					
Male					
Total					

4. Complete the above table for the 100 students who were surveyed.

5. The table shows the responses to the after-school activity question for males and females. Do you think there is a difference in the responses of males and females? Explain your answer.

Lesson 10: Summarizing Bivariate Categorical Data with Relative Frequencies

Classwork

This lesson expands on your work with two-way frequency tables from Lesson 9.

Exploratory Challenge 1: Extending the Frequency Table to a Relative Frequency Table

Determining the number of students in each cell presents the first step in organizing bivariate categorical data. Another way of analyzing the data in the table is to calculate the *relative frequency* for each cell. Relative frequencies relate each frequency count to the total number of observations. For each cell in this table, the *relative frequency* of a cell is found by dividing the frequency of that cell by the total number of responses.

Consider the two-way frequency table from the previous lesson.

Two-Way Frequency Table:

	To Fly	Freeze Time	Invisibility	Super Strength	Telepathy	Total
Females	49	60	48	1	70	228
Males	51	71	27	25	48	222
Total	100	131	75	26	118	450

The relative frequency table would be found by dividing each of the above cell values by 450. For example, the relative frequency of females selecting to fly is $\frac{49}{450}$, or approximately 0.109, to the nearest thousandth. A few of the other relative frequencies to the nearest thousandth are shown in the following relative frequency table:

	To Fly	Freeze Time	Invisibility	Super Strength	Telepathy	Total
Females	$\frac{49}{450} \approx 0.109$					$\frac{228}{450} \approx 0.507$
Males			$\frac{27}{450} \approx 0.060$			
Total		$\frac{131}{450} \approx 0.291$			$\frac{118}{450} \approx 0.262$	

Exercises 1–7

1. Calculate the remaining relative frequencies in the table below. Write the value in the table as a decimal rounded to the nearest thousandth or as a percent.

   Two-Way Frequency Table:

	To Fly	Freeze Time	Invisibility	Super Strength	Telepathy	Total
Females						
Males						
Total						

2. Based on previous work with frequency tables, which cells in this table would represent the joint relative frequencies?

3. Which cells in the relative frequency table would represent the marginal relative frequencies?

4. What is the joint relative frequency for females who selected invisibility as their favorite superpower?

5. What is the marginal relative frequency for freeze time? Interpret the meaning of this value.

Lesson 10:   Summarizing Bivariate Categorical Data with Relative Frequencies

EUREKA MATH™

6. What is the difference in the joint relative frequencies for males and for females who selected to fly as their favorite superpower?

7. Is there a noticeable difference between the genders and their favorite superpowers?

Exploratory Challenge 2: Interpreting Data

Interest in superheroes continues at Rufus King High School. The students who analyzed the data in the previous lesson decided to create a comic strip for the school website that involves a superhero. They thought the summaries developed from the data would be helpful in designing the comic strip.

Only one power will be given to the superhero. A debate arose as to what power the school's superhero would possess. Students used the two-way frequency table and the relative frequency table to continue the discussion. Take another look at those tables.

Scott initially indicated that the character created should have super strength as the special power. This suggestion was not well received by the other students planning this project. In particular, Jill argued, "Well, if you don't want to ignore more than half of the readers, then I suggest telepathy is the better power for our character."

Exercises 8–10

Scott acknowledged that super strength was probably not the best choice based on the data. "The data indicate that freeze time is the most popular power for a superhero," continued Scott. Jill, however, still did not agree with Scott that this was a good choice. She argued that telepathy was a better choice.

8. How do the data support Scott's claim? Why do you think he selected freeze time as the special power for the comic strip superhero?

9. How do the data support Jill's claim? Why do you think she selected telepathy as the special power for the comic strip superhero?

10. Of the two special powers freeze time and telepathy, select one and justify why you think it is a better choice based on the data.

Lesson Summary

- *Categorical data* are data that take on values that are categories rather than numbers. Examples include male or female for the categorical variable of gender or the five superpower categories for the categorical variable of superpower qualities.

- A *two-way frequency table* is used to summarize bivariate categorical data.

- A *relative frequency* compares a frequency count to the total number of observations. It can be written as a decimal or percent. A two-way table summarizing the relative frequencies of each cell is called a *relative frequency table*.

- The marginal cells in a two-way relative frequency table are called the *marginal relative frequencies*, while the joint cells are called the *joint relative frequencies*.

Problem Set

1. Consider the Rufus King High School data from the previous lesson regarding after-school activities:

	Intramural Basketball	Chess Club	Jazz Band	Not Involved	Total
Males	20	2	8	10	40
Females	20	10	10	20	60
Total	40	12	18	30	100

Calculate the relative frequencies for each of the cells to the nearest thousandth. Place the relative frequencies in the cells of the following table. (The first cell has been completed as an example.)

	Intramural Basketball	Chess Club	Jazz Band	Not Involved	Total
Males	$\dfrac{20}{100} = 0.200$				
Females					
Total					

2. Based on your relative frequency table, what is the relative frequency of students who indicated they play basketball?

3. Based on your table, what is the relative frequency of males who play basketball?

4. If a student were randomly selected from the students at the school, do you think the student selected would be a male or a female?

5. If a student were selected at random from school, do you think this student would be involved in an after-school program? Explain your answer.

6. Why might someone question whether or not the students who completed the survey were randomly selected? If the students completing the survey were randomly selected, what do the marginal relative frequencies possibly tell you about the school? Explain your answer.

7. Why might females think they are more involved in after-school activities than males? Explain your answer.

Lesson 11: Conditional Relative Frequencies and Association

Classwork

After further discussion, the students involved in designing the superhero comic strip decided that before any decision is made, a more careful look at the data on the special powers a superhero character could possess was needed. There is an association between gender and superpower response if the superpower responses of males are not the same as the superpower responses of females. Examining each row of the table can help determine whether or not there is an association.

Exploratory Challenge 1: Conditional Relative Frequencies

Recall the two-way table from the previous lesson.

	To Fly	Freeze Time	Invisibility	Super Strength	Telepathy	Total
Females	49	60	48	1	70	228
Males	51	71	27	25	48	222
Total	100	131	75	26	118	450

A *conditional relative frequency* compares a frequency count to the marginal total that represents the condition of interest. For example, the condition of interest in the first row is females. The row conditional relative frequency of females responding invisibility as the favorite superpower is $\frac{48}{228}$, or approximately 0.211. This conditional relative frequency indicates that approximately 21.1% of females prefer invisibility as their favorite superpower. Similarly, $\frac{27}{222}$, or approximately 0.122 or 12.2%, of males prefer invisibility as their favorite superpower.

Exercises 1–5

1. Use the frequency counts from the table in Exploratory Challenge 1 to calculate the missing row of conditional relative frequencies. Round the answers to the nearest thousandth.

	To Fly	Freeze Time	Invisibility	Super Strength	Telepathy	Total
Females			$\frac{48}{228} \approx 0.211$			
Males	$\frac{51}{222} \approx 0.230$					$\frac{222}{222} = 1.000$
Total						

2. Suppose that a student is selected at random from those who completed the survey. What do you think is the gender of the student selected? What would you predict for this student's response to the superpower question?

3. Suppose that a student is selected at random from those who completed the survey. If the selected student is male, what do you think was his response to the selection of a favorite superpower? Explain your answer.

4. Suppose that a student is selected at random from those who completed the survey. If the selected student is female, what do you think was her response to the selection of a favorite superpower? Explain your answer.

5. What superpower was selected by approximately one-third of the females? What superpower was selected by approximately one-third of the males? How did you determine each answer from the conditional relative frequency table?

Exploratory Challenge 2: Possible Association Based on Conditional Relative Frequencies

Two categorical variables are associated if the row conditional relative frequencies (or column relative frequencies) are different for the rows (or columns) of the table. For example, if the selection of superpowers selected for females is different than the selection of superpowers for males, then gender and superpower favorites are associated. This difference indicates that knowing the gender of a person in the sample indicates something about their superpower preference.

The evidence of an association is strongest when the conditional relative frequencies are quite different. If the conditional relative frequencies are nearly equal for all categories, then there is probably not an association between variables.

Exercises 6–10

Examine the conditional relative frequencies in the two-way table of conditional relative frequencies you created in Exercise 1. Note that for each superpower, the conditional relative frequencies are different for females and males.

6. For what superpowers would you say that the conditional relative frequencies for females and males are very different?

7. For what superpowers are the conditional relative frequencies nearly equal for males and females?

8. Suppose a student is selected at random from the students who completed the survey. Would knowing the student's gender be helpful in predicting which superpower this student selected? Explain your answer.

9. Is there evidence of an association between gender and a favorite superpower? Explain why or why not.

10. What superpower would you recommend the students at Rufus King High School select for their superhero character? Justify your choice.

Exploratory Challenge 3: Association and Cause-and-Effect

Students were given the opportunity to prepare for a college placement test in mathematics by taking a review course. Not all students took advantage of this opportunity. The following results were obtained from a random sample of students who took the placement test.

	Placed in Math 200	Placed in Math 100	Placed in Math 50	Total
Took Review Course	40	13	7	60
Did Not Take Review Course	10	15	15	40
Total	50	28	22	100

Exercises 11–16

11. Construct a row conditional relative frequency table of the above data.

	Placed in Math 200	Placed in Math 100	Placed in Math 50	Total
Took Review Course				
Did Not Take Review Course				
Total				

12. Based on the conditional relative frequencies, is there evidence of an association between whether a student takes the review course and the math course in which the student was placed? Explain your answer.

13. Looking at the conditional relative frequencies, the proportion of students who placed into Math 200 is much higher for those who took the review course than for those who did not. One possible explanation is that taking the review course caused improvement in placement test scores. What is another possible explanation?

Now consider the following statistical study:

Fifty students were selected at random from students at a large middle school. Each of these students was classified according to sugar consumption (high or low) and exercise level (high or low). The resulting data are summarized in the following frequency table.

		Exercise Level		
		High	**Low**	**Total**
Sugar Consumption	**High**	14	18	32
	Low	14	4	18
	Total	28	22	50

14. Calculate the row conditional relative frequencies, and display them in a row conditional relative frequency table.

		Exercise Level		
		High	Low	Total
Sugar Consumption	High			
	Low			
	Total			

15. Is there evidence of an association between sugar consumption category and exercise level? Support your answer using conditional relative frequencies.

16. Is it reasonable to conclude that high sugar consumption is the cause of the observed differences in the conditional relative frequencies? What other explanations could explain a difference in the conditional relative frequencies? Explain your answer.

Lesson Summary

- A conditional relative frequency compares a frequency count to the marginal total that represents the *condition* of interest.

- The differences in conditional relative frequencies are used to assess whether or not there is an association between two categorical variables.

- The greater the differences in the conditional relative frequencies, the stronger the evidence that an association exits.

- An observed association between two variables does not necessarily mean that there is a cause-and-effect relationship between the two variables.

Problem Set

Consider again the summary of data from the 100 randomly selected students in the Rufus King High School investigation of after-school activities and gender.

	Intramural Basketball	Chess Club	Jazz Band	Not Involved	Total
Females	20	10	10	20	60
Males	20	2	8	10	40
Total	40	12	18	30	100

1. Construct a row conditional relative frequency table for this data. Decimal values are given to the nearest thousandth.

	Intramural Basketball	Chess Club	Jazz Band	Not Involved	Total
Females					60
Males					40
Total					

2. For what after-school activities do you think the row conditional relative frequencies for females and males are very different? What might explain why males or females select different activities?

3. If John, a male student at Rufus King High School, completed the after-school survey, what would you predict was his response? Explain your answer.

Lesson 11:   Conditional Relative Frequencies and Association

4. If Beth, a female student at Rufus King High School, completed the after-school survey, what would you predict was her response? Explain your answer.

5. Notice that 20 female students participate in intramural basketball and that 20 male students participate in intramural basketball. Is it accurate to say that females and males are equally involved in intramural basketball? Explain your answer.

6. Do you think there is an association between gender and choice of after-school program? Explain.

Column conditional relative frequencies can also be computed by dividing each frequency in a frequency table by the corresponding column total to create a column conditional relative frequency table. Column conditional relative frequencies indicate the proportions, or relative frequencies, based on the column totals.

7. If you wanted to know the relative frequency of females surveyed who participated in chess club, would you use a row conditional relative frequency or a column conditional relative frequency?

8. If you wanted to know the relative frequency of band members surveyed who were female, would you use a row conditional relative frequency or a column conditional relative frequency?

9. For the superpower survey data, write a question that would be answered using a row conditional relative frequency.

10. For the superpower survey data, write a question that would be answered using a column conditional relative frequency.

This page intentionally left blank

Lesson 12: Relationships Between Two Numerical Variables

Classwork

A scatter plot is an informative way to display numerical data with two variables. In your previous work in Grade 8, you saw how to construct and interpret scatter plots. Recall that if the two numerical variables are denoted by x and y, the scatter plot of the data is a plot of the (x, y) data pairs.

Example 1: Looking for Patterns in a Scatter Plot

The National Climate Data Center collects data on weather conditions at various locations. They classify each day as clear, partly cloudy, or cloudy. Using data taken over a number of years, they provide data on the following variables.

x represents elevation above sea level (in feet).

y represents mean number of clear days per year.

w represents mean number of partly cloudy days per year.

z represents mean number of cloudy days per year.

The table below shows data for 14 U.S. cities.

City	x (Elevation Above Sea Level in Feet)	y (Mean Number of Clear Days per Year)	w (Mean Number of Partly Cloudy Days per Year)	z (Mean Number of Cloudy Days per Year)
Albany, NY	275	69	111	185
Albuquerque, NM	5,311	167	111	87
Anchorage, AK	114	40	60	265
Boise, ID	2,838	120	90	155
Boston, MA	15	98	103	164
Helena, MT	3,828	82	104	179
Lander, WY	5,557	114	122	129
Milwaukee, WI	672	90	100	175
New Orleans, LA	4	101	118	146
Raleigh, NC	434	111	106	149
Rapid City, SD	3,162	111	115	139
Salt Lake City, UT	4,221	125	101	139
Spokane, WA	2,356	86	88	191
Tampa, FL	19	101	143	121

Here is a scatter plot of the data on elevation and mean number of clear days.

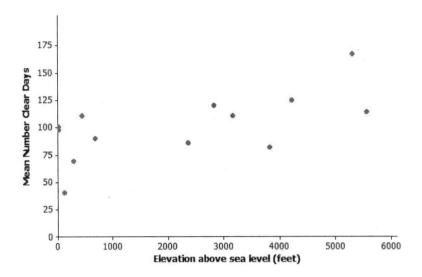

Data Source: www.ncdc.noaa.gov

Exercises 1–3

1. Do you see a pattern in the scatter plot, or does it look like the data points are scattered?

2. How would you describe the relationship between elevation and mean number of clear days for these 14 cities? That is, does the mean number of clear days tend to increase as elevation increases, or does the mean number of clear days tend to decrease as elevation increases?

3. Do you think that a straight line would be a good way to describe the relationship between the mean number of clear days and elevation? Why do you think this?

EUREKA
MATH™

Exercises 4–7: Thinking about Linear Relationships

Below are three scatter plots. Each one represents a data set with eight observations.

The scales on the x- and y-axes have been left off these plots on purpose, so you have to think carefully about the relationships.

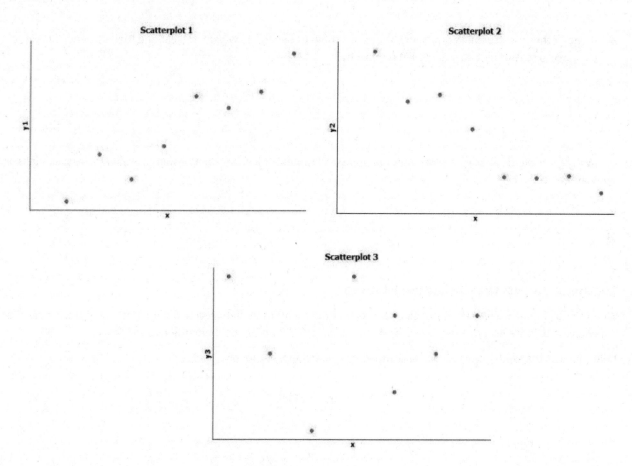

4. If one of these scatter plots represents the relationship between height and weight for eight adults, which scatter plot do you think it is and why?

5. If one of these scatter plots represents the relationship between height and SAT math score for eight high school seniors, which scatter plot do you think it is and why?

6. If one of these scatter plots represents the relationship between the weight of a car and fuel efficiency for eight cars, which scatter plot do you think it is and why?

7. Which of these three scatter plots does *not* appear to represent a linear relationship? Explain the reasoning behind your choice.

Exercises 8–13: Not Every Relationship Is Linear

When a straight line provides a reasonable summary of the relationship between two numerical variables, we say that the two variables are *linearly related* or that there is a *linear relationship* between the two variables.

Take a look at the scatter plots below, and answer the questions that follow.

Scatter Plot 1

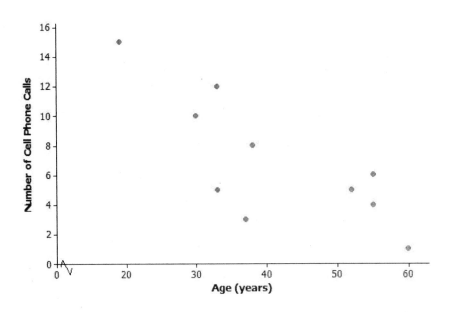

EUREKA
MATH™

8. Is there a relationship between the number of cell phone calls and age, or does it look like the data points are scattered?

9. If there is a relationship between the number of cell phone calls and age, does the relationship appear to be linear?

Scatter Plot 2

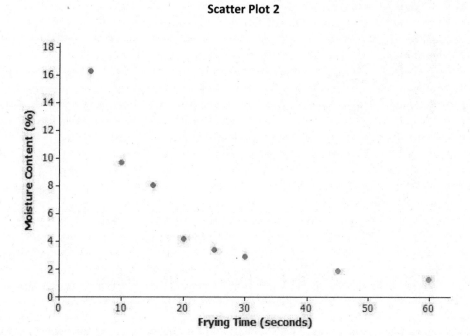

Data Source: R.G. Moreira, J. Palau, V.E. Sweat, and X. Sun, "Thermal and Physical Properties of Tortilla Chips as a Function of Frying Time," *Journal of Food Processing and Preservation,* 19 (1995): 175.

10. Is there a relationship between moisture content and frying time, or do the data points look scattered?

11. If there is a relationship between moisture content and frying time, does the relationship look linear?

Scatter Plot 3

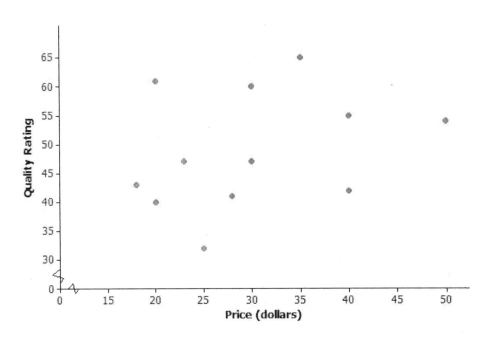

Data Source: www.consumerreports.org/health

12. Scatter Plot 3 shows data for the prices of bike helmets and the quality ratings of the helmets (based on a scale that estimates helmet quality). Is there a relationship between quality rating and price, or are the data points scattered?

13. If there is a relationship between quality rating and price for bike helmets, does the relationship appear to be linear?

Lesson 12:   Relationships Between Two Numerical Variables

EUREKA MATH™

Lesson Summary

- A scatter plot can be used to investigate whether or not there is a relationship between two numerical variables.

- A relationship between two numerical variables can be described as a linear or nonlinear relationship.

Problem Set

1. Construct a scatter plot that displays the data for x (elevation above sea level in feet) and w (mean number of *partly cloudy days per year*).

City	x (Elevation Above Sea Level in Feet)	y (Mean Number of Clear Days per Year)	w (Mean Number of Partly Cloudy Days per Year)	z (Mean Number of Cloudy Days per Year)
Albany, NY	275	69	111	185
Albuquerque, NM	5,311	167	111	87
Anchorage, AK	114	40	60	265
Boise, ID	2,838	120	90	155
Boston, MA	15	98	103	164
Helena, MT	3,828	82	104	179
Lander, WY	5,557	114	122	129
Milwaukee, WI	672	90	100	175
New Orleans, LA	4	101	118	146
Raleigh, NC	434	111	106	149
Rapid City, SD	3,162	111	115	139
Salt Lake City, UT	4,221	125	101	139
Spokane, WA	2,356	86	88	191
Tampa, FL	19	101	143	121

2. Based on the scatter plot you constructed in Problem 1, is there a relationship between elevation and the mean number of partly cloudy days per year? If so, how would you describe the relationship? Explain your reasoning.

Consider the following scatter plot for Problems 3 and 4.

Scatter Plot 4

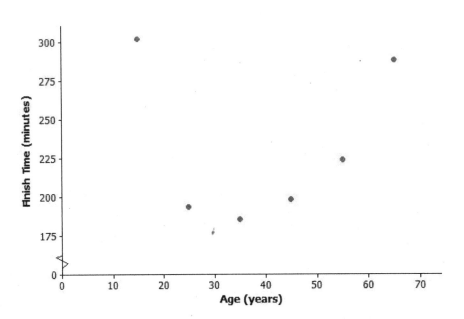

Data Source: Sample of six women who ran the 2003 NYC marathon

3. Is there a relationship between finish time and age, or are the data points scattered?

4. Do you think there is a relationship between finish time and age? If so, does it look linear?

Lesson 12:   Relationships Between Two Numerical Variables

EUREKA
MATH™

Consider the following scatter plot for Problems 5 and 6.

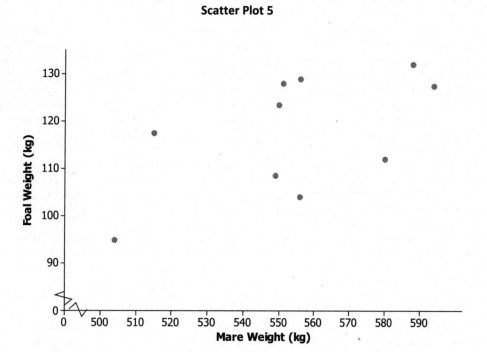

Scatter Plot 5

Data Source: Elissa Z. Cameron, Kevin J. Stafford, Wayne L. Linklater, and Clare J. Veltman, "Suckling behaviour does not measure milk intake in horses, equus caballus," *Animal Behaviour,* 57 (1999): 673.

5. A mare is a female horse, and a foal is a baby horse. Is there a relationship between a foal's birth weight and a mare's weight, or are the data points scattered?

6. If there is a relationship between baby birth weight and mother's weight, does the relationship look linear?

This page intentionally left blank

Lesson 13: Relationships Between Two Numerical Variables

Classwork

Not all relationships between two numerical variables are *linear*. There are many situations where the pattern in the scatter plot would best be described by a curve. Two types of functions often used in modeling nonlinear relationships are *quadratic* and *exponential* functions.

Example 1: Modeling Relationships

Sometimes the pattern in a scatter plot looks like the graph of a quadratic function (with the points falling roughly in the shape of a *U* that opens up or down), as in the graph below.

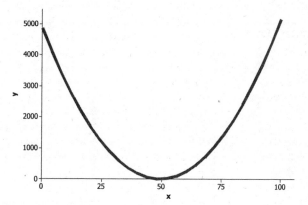

In other situations, the pattern in the scatter plot might look like the graphs of exponential functions that either are upward sloping (Graph 1) or downward sloping (Graph 2).

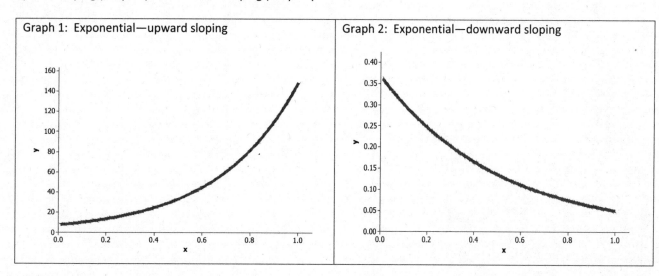

Exercises 1–6

Consider again the five scatter plots discussed in the previous lesson.

Scatter Plot 1

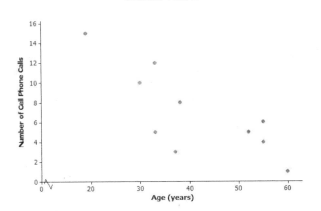

Scatter Plot 2

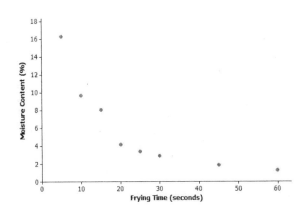

Scatter Plot 3

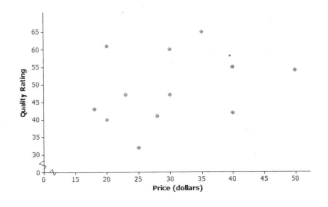

Scatter Plot 4

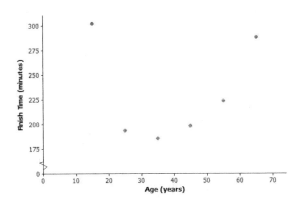

Scatter Plot 5

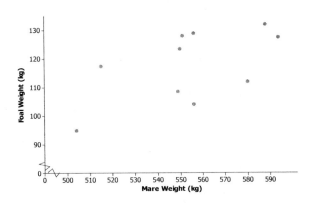

Lesson 13:   Relationships Between Two Numerical Variables

EUREKA
MATH™

1. Which of the five scatter plots from Lesson 12 shows a pattern that could be reasonably described by a quadratic curve?

2. Which of the five scatter plots shows a pattern that could be reasonably described by an exponential curve?

Let's revisit the data on elevation (in feet above sea level) and mean number of clear days per year. The scatter plot of this data is shown below. The plot also shows a straight line that can be used to model the relationship between elevation and mean number of clear days. (In Grade 8, you informally fit a straight line to model the relationship between two variables. The next lesson shows a more formal way to fit a straight line.) The equation of this line is $y = 83.6 + 0.008x$.

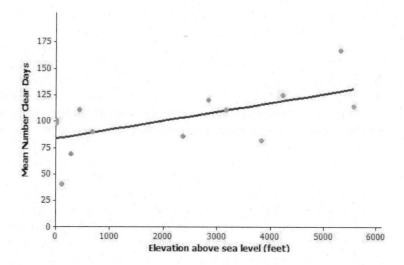

3. Assuming that the 14 cities used in this scatter plot are representative of cities across the United States, should you see more clear days per year in Los Angeles, which is near sea level, or in Denver, which is known as the mile-high city? Justify your choice with a line showing the relationship between elevation and mean number of clear days.

4. One of the cities in the data set was Albany, New York, which has an elevation of 275 ft. If you did not know the mean number of clear days for Albany, what would you predict this number to be based on the line that describes the relationship between elevation and mean number of clear days?

5. Another city in the data set was Albuquerque, New Mexico. Albuquerque has an elevation of 5,311 ft. If you did not know the mean number of clear days for Albuquerque, what would you predict this number to be based on the line that describes the relationship between elevation and mean number of clear days?

6. Was the prediction of the mean number of clear days based on the line closer to the actual value for Albany with 69 clear days or for Albuquerque with 167 clear days? How could you tell this from looking at the scatter plot with the line shown above?

Example 2: A Quadratic Model

Farmers sometimes use fertilizers to increase crop yield but often wonder just how much fertilizer they should use. The data shown in the scatter plot below are from a study of the effect of fertilizer on the yield of corn.

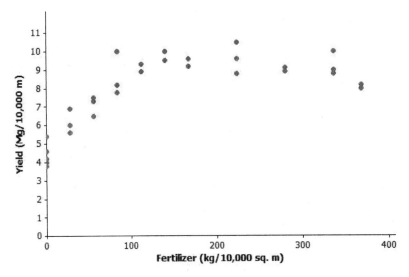

Data Source: M.E. Cerrato and A.M. Blackmer, "Comparison of Models for Describing Corn Yield Response to Nitrogen Fertilizer" *Agronomy Journal,* 82 (1990): 138.

EUREKA
MATH™

Exercises 7–9

7. The researchers who conducted this study decided to use a quadratic curve to describe the relationship between yield and amount of fertilizer. Explain why they made this choice.

8. The model that the researchers used to describe the relationship was $y = 4.7 + 0.05x - 0.0001x^2$, where x represents the amount of fertilizer (kg per 10,000 sq. m) and y represents corn yield (Mg per 10,000 sq. m). Use this quadratic model to complete the following table. Then sketch the graph of this quadratic equation on the scatter plot.

x	y
0	
100	
200	
300	
400	

9. Based on this quadratic model, how much fertilizer per 10,000 sq. m would you recommend that a farmer use on his cornfields in order to maximize crop yield? Justify your choice.

Example 3: An Exponential Model

How do you tell how old a lobster is? This question is important to biologists and to those who regulate lobster trapping. To answer this question, researchers recorded data on the shell length of 27 lobsters that were raised in a laboratory and whose ages were known.

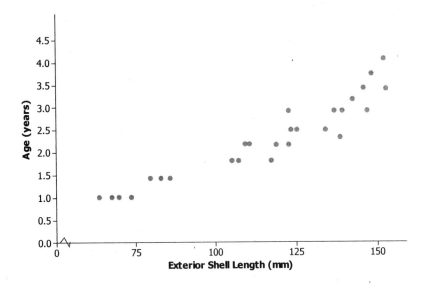

Data Source: Kerry E. Maxwell, Thomas R. Matthews, Matt R.J. Sheehy, Rodney D. Bertelsen, and Charles D. Derby, "Neurolipofuscin is a Measure of Age in *Panulirus argus*, the Caribbean Spiny Lobster, in Florida" *Biological Bulletin,* 213 (2007): 55.

Exercises 10–13

10. The researchers who conducted this study decided to use an exponential curve to describe the relationship between age and exterior shell length. Explain why they made this choice.

Lesson 13:   Relationships Between Two Numerical Variables

11. The model that the researchers used to describe the relationship is $y = 10^{-0.403 + 0.0063x}$, where x represents the exterior shell length (mm), and y represents the age of the lobster (in years). The exponential curve is shown on the scatter plot below. Does this model provide a good description of the relationship between age and exterior shell length? Explain why or why not.

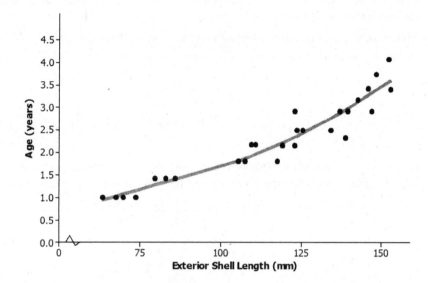

12. Based on this exponential model, what age is a lobster with an exterior shell length of 100 mm?

13. Suppose that trapping regulations require that any lobster with an exterior shell length less than 75 mm or more than 150 mm must be released. Based on the exponential model, what are the ages of lobsters with exterior shell lengths less than 75 mm? What are the ages of lobsters with exterior shell lengths greater than 150 mm? Explain how you arrived at your answer.

Lesson Summary

- A scatter plot can be used to investigate whether or not there is a relationship between two numerical variables.
- Linear, quadratic, and exponential functions are common models that can be used to describe the relationship between variables.
- Models can be used to answer questions about how two variables are related.

Problem Set

Biologists conducted a study of the nesting behavior of a type of bird called a flycatcher. They examined a large number of nests and recorded the latitude for the location of the nest and the number of chicks in the nest.

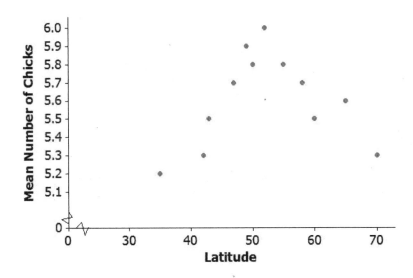

Data Source: Juan José Sanz, "Geographic variation in breeding parameters of the pied flycatcher *Ficedula hypoleuca*"
Ibis, 139 (1997): 107.

1. What type of model (linear, quadratic, or exponential) would best describe the relationship between latitude and mean number of chicks?

2. One model that could be used to describe the relationship between mean number of chicks and latitude is $y = 0.175 + 0.21x - 0.002x^2$, where x represents the latitude of the location of the nest and y represents the number of chicks in the nest. Use the quadratic model to complete the following table. Then sketch a graph of the quadratic curve on the scatter plot provided at the beginning of the Problem Set.

x (degrees)	y
30	
40	
50	
60	
70	

3. Based on this quadratic model, what is the best latitude for hatching the most flycatcher chicks? Justify your choice.

Suppose that social scientists conducted a study of senior citizens to see how the time (in minutes) required to solve a word puzzle changes with age. The scatter plot below displays data from this study.

Let x equal the age of the citizen and y equal the time (in minutes) required to solve a word puzzle for the seven study participants.

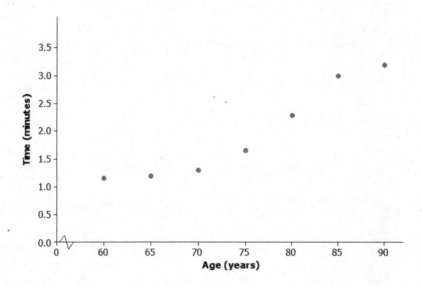

4. What type of model (linear, quadratic, or exponential) would you use to describe the relationship between age and time required to complete the word puzzle?

5. One model that could describe the relationship between age and time to complete the word puzzle is $y = 10^{-1.01\,+\,0.017x}$. This exponential curve is shown on the scatter plot below. Does this model do a good job of describing the relationship between age and time to complete the word puzzle? Explain why or why not.

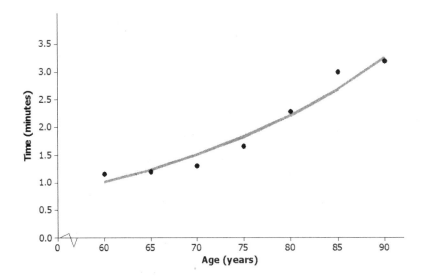

6. Based on this exponential model, what time would you predict for a person who is 78 years old?

EUREKA
MATH™

Lesson 14: Modeling Relationships with a Line

Classwork

Example 1: Using a Line to Describe a Relationship

Kendra likes to watch crime scene investigation shows on television. She watched a show where investigators used a shoe print to help identify a suspect in a case. She questioned how possible it is to predict someone's height from his shoe print.

To investigate, she collected data on shoe length (in inches) and height (in inches) from 10 adult men. Her data appear in the table and scatter plot below.

x (Shoe Length)	y (Height)
12.6	74
11.8	65
12.2	71
11.6	67
12.2	69
11.4	68
12.8	70
12.2	69
12.6	72
11.8	71

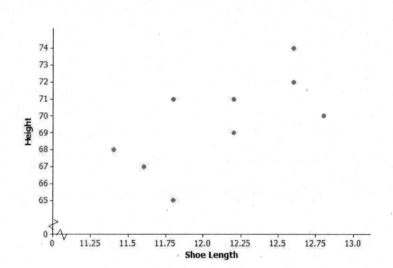

Exercises 1–2

1. Is there a relationship between shoe length and height?

2. How would you describe the relationship? Do the men with longer shoe lengths tend be taller?

Example 2: Using Models to Make Predictions

When two variables x and y are linearly related, you can use a line to describe their relationship. You can also use the equation of the line to predict the value of the y-variable based on the value of the x-variable.

For example, the line $y = 25.3 + 3.66x$ might be used to describe the relationship between shoe length and height, where x represents shoe length and y represents height. To predict the height of a man with a shoe length of 12 in., you would substitute 12 for x in the equation of the line and then calculate the value of y.

$$y = 25.3 + 3.66x = 25.3 + 3.66(12) = 69.22$$

You would predict a height of 69.22 in. for a man with a shoe length of 12 in.

Exercises 3–7

3. Below is a scatter plot of the data with two linear models, $y = 130 - 5x$ and $y = 25.3 + 3.66x$. Which of these two models does a better job of describing how shoe length (x) and height (y) are related? Explain your choice.

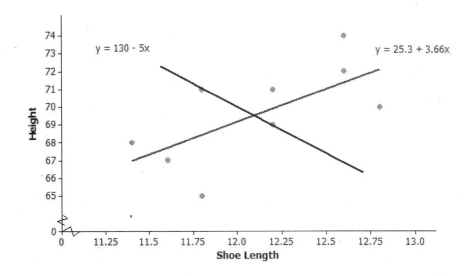

4. One of the men in the sample has a shoe length of 11.8 in. and a height of 71 in. Circle the point in the scatter plot in Exercise 3 that represents this man.

Lesson 14:   Modeling Relationships with a Line

EUREKA MATH™

5. Suppose that you do not know this man's height but do know that his shoe length is 11.8 in. If you use the model $y = 25.3 + 3.66x$, what would you predict his height to be? If you use the model $y = 130 - 5x$, what would you predict his height to be?

6. Which model was closer to the actual height of 71 in.? Is that model a better fit to the data? Explain your answer.

7. Is there a better way to decide which of two lines provides a better description of a relationship (rather than just comparing the predicted value to the actual value for one data point in the sample)?

Example 3: Residuals

One way to think about how useful a line is for describing a relationship between two variables is to use the line to predict the y-values for the points in the scatter plot. These predicted values could then be compared to the actual y-values.

For example, the first data point in the table represents a man with a shoe length of 12.6 in. and height of 74 in. If you use the line $y = 25.3 + 3.66x$ to predict this man's height, you would get:

$$y = 25.3 + 3.66x$$
$$= 25.3 + 3.66(12.6)$$
$$= 71.42$$

His predicted height is 71.42 in. Because his actual height was 74 in., you can calculate the prediction error by subtracting the predicted value from the actual value. This prediction error is called a *residual*. For the first data point, the residual is calculated as follows:

$$\text{Residual} = \text{actual } y\text{-value} - \text{predicted } y\text{-value}$$
$$= 74 - 71.42$$
$$= 2.58$$

Exercises 8–10

8. For the line $y = 25.3 + 3.66x$, calculate the missing values, and add them to complete the table.

x (Shoe Length)	y (Height)	Predicted y-value	Residual
12.6	74	71.42	2.58
11.8	65		−3.49
12.2	71		
11.6	67	67.76	−0.76
12.2	69	69.95	−0.95
11.4	68	67.02	
12.8	70	72.15	−2.15
12.2	69		−0.95
12.6	72	71.42	0.58
11.8	71	68.49	2.51

9. Why is the residual in the table's first row positive and the residual in the second row negative?

10. What is the sum of the residuals? Why did you get a number close to zero for this sum? Does this mean that all of the residuals were close to 0?

Exercises 11–13

When you use a line to describe the relationship between two numerical variables, the *best* line is the line that makes the residuals as small as possible overall.

11. If the residuals tend to be small, what does that say about the fit of the line to the data?

Lesson 14:   Modeling Relationships with a Line

The most common choice for the *best* line is the line that makes the sum of the *squared* residuals as small as possible. Add a column on the right of the table in Exercise 8. Calculate the square of each residual and place the answer in the column.

12. Why do we use the sum of the squared residuals instead of just the sum of the residuals (without squaring)? Hint: Think about whether the sum of the residuals for a line can be small even if the prediction errors are large. Can this happen for squared residuals?

13. What is the sum of the squared residuals for the line $y = 25.3 + 3.66x$ and the data of Exercise 11?

Example 4: The Least Squares Line (Best-Fit Line)

The line that has a smaller sum of squared residuals for this data set than any other line is called the *least squares line*. This line can also be called the *best-fit line* or the *line of best fit* (or regression line).

For the shoe-length and height data for the sample of 10 men, the line $y = 25.3 + 3.66x$ is the least squares line. No other line would have a smaller sum of squared residuals for this data set than this line.

There are equations that can be used to calculate the value for the slope and the intercept of the least squares line, but these formulas require a lot of tedious calculations. Fortunately, a graphing calculator can be used to find the equation of the least squares line.

Your teacher will show you how to enter data and obtain the equation of the least squares line using your graphing calculator or other statistics program.

Exercises 14–17

14. Enter the shoe-length and height data, and then use your calculator to find the equation of the least squares line. Did you get $y = 25.3 + 3.66x$? (The slope and y-intercept here have been rounded to the nearest hundredth.)

15. Assuming that the 10 men in the sample are representative of adult men in general, what height would you predict for a man whose shoe length is 12.5 in.? What height would you predict for a man whose shoe length is 11.9 in.?

Once you have found the equation of the least squares line, the values of the slope and y-intercept of the line often reveal something interesting about the relationship you are modeling.

The slope of the least squares line is the change in the predicted value of the y-variable associated with an increase of one in the value of the x-variable.

16. Give an interpretation of the slope of the least squares line $y = 25.3 + 3.66x$ for predicting height from shoe size for adult men.

The y-intercept of a line is the predicted value of y when x equals zero. When using a line as a model for the relationship between two numerical variables, it often does not make sense to interpret the y-intercept because a x-value of zero may not make any sense.

17. Explain why it does not make sense to interpret the y-intercept of 25.3 as the predicted height for an adult male whose shoe length is zero.

> **Lesson Summary**
>
> When the relationship between two numerical variables x and y is linear, a straight line can be used to describe the relationship. Such a line can then be used to predict the value of y based on the value of x. When a prediction is made, the prediction error is the difference between the actual y-value and the predicted y-value. The prediction error is called a *residual*, and the residual is calculated as *residual = actual y-value − predicted y-value*. The *least squares line* is the line that is used to model a linear relationship. The least squares line is the *best* line in that it has a smaller sum of squared residuals than any other line.

Problem Set

Kendra wondered if the relationship between shoe length and height might be different for men and women. To investigate, she also collected data on shoe length (in inches) and height (in inches) for 12 women.

x (Shoe Length of Women)	y (Height of Women)
8.9	61
9.6	61
9.8	66
10.0	64
10.2	64
10.4	65
10.6	65
10.6	67
10.5	66
10.8	67
11.0	67
11.8	70

1. Construct a scatter plot of these data.

2. Is there a relationship between shoe length and height for these 12 women?

3. Find the equation of the least squares line. (Round values to the nearest hundredth.)

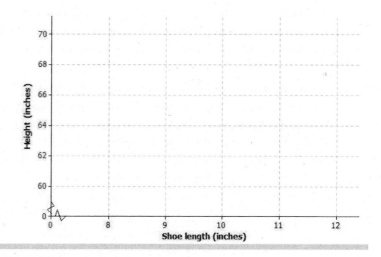

EUREKA
MATH™

Lesson 14: Modeling Relationships with a Line

S.107

This work is derived from Eureka Math ™ and licensed by Great Minds. ©2015 Great Minds. eureka-math.org
ALG1-M2-SE-B1-1.3.0-05.2015

4. Suppose that these 12 women are representative of adult women in general. Based on the least squares line, what would you predict for the height of a woman whose shoe length is 10.5 in.? What would you predict for the height of a woman whose shoe length is 11.5 in.?

5. One of the women in the sample had a shoe length of 9.8 in. Based on the regression line, what would you predict for her height?

6. What is the value of the residual associated with the observation for the woman with the shoe length of 9.8 in.?

7. Add the predicted value and the residual you just calculated to the table below. Then, calculate the sum of the squared residuals.

x (Shoe Length of Women)	y (Height of Women)	Predicted Height (in.)	Residual (in.)	Squared Residual
8.9	61	60.72	0.28	
9.6	61	62.92	−1.92	
9.8	66			
10.0	64	64.18	−0.18	
10.2	64	64.81	−0.81	
10.4	65	65.44	−0.44	
10.6	65	66.07	−1.07	
10.6	67	66.07	0.93	
10.5	66	65.76	0.24	
10.8	67	66.7	0.3	
11.0	67	67.33	−0.33	
11.8	70	69.85	0.15	

8. Provide an interpretation of the slope of the least squares line.

9. Does it make sense to interpret the y-intercept of the least squares line in this context? Explain why or why not.

10. Would the sum of the squared residuals for the line $y = 25 + 2.8x$ be greater than, about the same as, or less than the sum you computed in Problem 7? Explain how you know this. You should be able to answer this question without calculating the sum of squared residuals for this new line.

11. For the men, the least squares line that describes the relationship between x, which represents shoe length (in inches), and y, which represents height (in inches), was $y = 25.3 + 3.66x$. How does this compare to the equation of the least squares line for women? Would you use $y = 25.3 + 3.66x$ to predict the height of a woman based on her shoe length? Explain why or why not.

12. Below are dot plots of the shoe lengths for women and the shoe lengths for men. Suppose that you found a shoe print and that when you measured the shoe length, you got 10.8 in. Do you think that a man or a woman left this shoe print? Explain your choice.

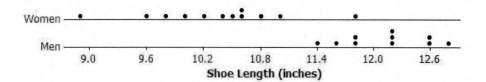

13. Suppose that you find a shoe print and the shoe length for this print is 12 in. What would you predict for the height of the person who left this print? Explain how you arrived at this answer.

This page intentionally left blank

Lesson 15: Interpreting Residuals from a Line

Classwork

Example 1: Calculating Prediction Errors

The gestation time for an animal is the typical duration between conception and birth. The longevity of an animal is the typical lifespan for that animal. The gestation times (in days) and longevities (in years) for 13 types of animals are shown in the table below.

Animal	Gestation Time (days)	Longevity (years)
Baboon	187	20
Black Bear	219	18
Beaver	105	5
Bison	285	15
Cat	63	12
Chimpanzee	230	20
Cow	284	15
Dog	61	12
Fox (Red)	52	7
Goat	151	8
Lion	100	15
Sheep	154	12
Wolf	63	5

Data Source: *Core Math Tools,* http://nctm.org

Here is the scatter plot for this data set:

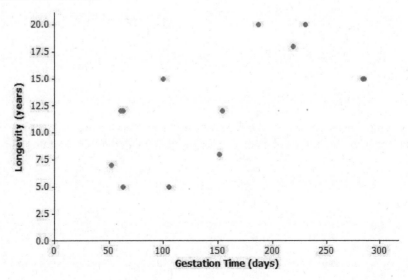

Exercises 1–4

Finding the equation of the least squares line relating longevity to gestation time for these types of animals provides the equation to predict longevity. How good is the line? In other words, if you were given the gestation time for another type of animal not included in the original list, how accurate would the least squares line be at predicting the longevity of that type of animal?

1. Using a graphing calculator, verify that the equation of the least squares line is $y = 6.642 + 0.03974x$, where x represents the gestation time (in days), and y represents longevity (in years).

   The least squares line has been added to the scatter plot below.

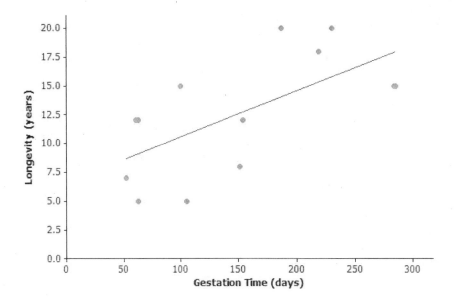

2. Suppose a particular type of animal has a gestation time of 200 days. Approximately what value does the line predict for the longevity of that type of animal?

3. Would the value you predicted in Exercise 2 necessarily be the exact value for the longevity of that type of animal? Could the actual longevity of that type of animal be longer than predicted? Could it be shorter?

EUREKA
MATH™

You can investigate further by looking at the types of animals included in the original data set. Take the lion, for example. Its gestation time is 100 days. You also know that its longevity is 15 years, but what does the least squares line *predict* for the lion's longevity?

Substituting $x = 100$ days into the equation, you get $y = 6.642 + 0.03974(100)$ or approximately 10.6. The least squares line predicts the lion's longevity to be approximately 10.6 years.

4. How close is this to being correct? More precisely, how much do you have to add to 10.6 to get the lion's true longevity of 15?

You can show the prediction error of 4.4 years on the graph like this:

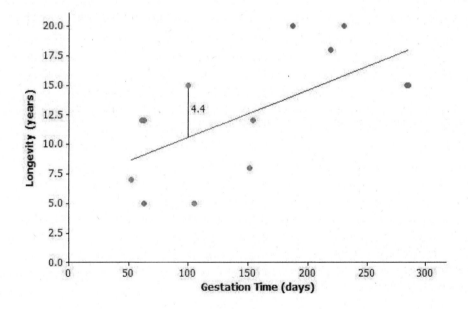

Exercises 5–6

5. Let's continue to think about the gestation times and longevities of animals. Let's specifically investigate how accurately the least squares line predicted the longevity of the black bear.

   a. What is the gestation time for the black bear?

---

b. Look at the graph. Roughly what does the least squares line predict for the longevity of the black bear?

c. Use the gestation time from part (a) and the least squares line $y = 6.642 + 0.03974x$ to predict the black bear's longevity. Round your answer to the nearest tenth.

d. What is the actual longevity of the black bear?

e. How much do you have to add to the predicted value to get the actual longevity of the black bear?

f. Show your answer to part (e) on the graph as a vertical line segment.

6. Repeat this activity for the sheep.

a. Substitute the sheep's gestation time for x into the equation to find the predicted value for the sheep's longevity. Round your answer to the nearest tenth.

b. What do you have to add to the predicted value in order to get the actual value of the sheep's longevity? (Hint: Your answer should be negative.)

Lesson 15:   Interpreting Residuals from a Line

c. Show your answer to part (b) on the graph as a vertical line segment. Write a sentence describing points in the graph for which a negative number would need to be added to the predicted value in order to get the actual value.

Example 2: Residuals as Prediction Errors

In previous exercises, you found out how much needs to be added to the predicted value to find the true value of an animal's longevity. In order to find this, you have been calculating

$$\text{actual value} - \text{predicted value.}$$

This quantity is referred to as a residual. It is summarized as

$$\text{residual} = \text{actual } y\text{-value} - \text{predicted } y\text{-value.}$$

You can now work out the residuals for all of the points in our animal longevity example. The values of the residuals are shown in the table below.

Animal	Gestation Time (days)	Longevity (years)	Residual (years)
Baboon	187	20	5.9
Black Bear	219	18	2.7
Beaver	105	5	−5.8
Bison	285	15	−3.0
Cat	63	12	2.9
Chimpanzee	230	20	4.2
Cow	284	15	−2.9
Dog	61	12	2.9
Fox (Red)	52	7	−1.7
Goat	151	8	−4.6
Lion	100	15	4.4
Sheep	154	12	−0.8
Wolf	63	5	−4.1

These residuals show that the actual longevity of an animal should be within six years of the longevity predicted by the least squares line.

Suppose you selected a type of animal that is not included in the original data set, and the gestation time for this type of animal is 270 days. Substituting $x = 270$ into the equation of the least squares line you get

$$y = 6.642 + 0.03974(270)$$
$$= 17.4.$$

The predicted longevity of this animal is 17.4 years.

Exercises 7–8

Think about what the *actual* longevity of this type of animal might be.

7. Could it be 30 years? How about 5 years?

8. Judging by the size of the residuals in our table, what kind of values do you think would be reasonable for the longevity of this type of animal?

Exercises 9–10

Continue to think about the gestation times and longevities of animals. The gestation time for the type of animal called the ocelot is known to be 85 days.

The least squares line predicts the longevity of the ocelot to be 10.0 years.

$$y = 6.642 + 0.03974(85) = 10.0$$

9. Based on the residuals in Example 3, would you be surprised to find that the longevity of the ocelot was 2 years? Why or why not? What might be a sensible range of values for the actual longevity of the ocelot?

10. We know that the actual longevity of the ocelot is 9 years. What is the residual for the ocelot?

This work is derived from Eureka Math ™ and licensed by Great Minds. ©2015 Great Minds. eureka-math.org
ALG1-M2-SE-B1-1.3.0-05.2015

Lesson Summary

- When a least squares line is used to calculate a predicted value, the prediction error can be measured by

  residual = actual y-value − predicted y-value.

- On the graph, the residuals are the vertical distances of the points from the least squares line.

- The residuals give us an idea how close a prediction might be when the least squares line is used to make a prediction for a value that is not included in the data set.

Problem Set

The time spent in surgery and the cost of surgery was recorded for six patients. The results and scatter plot are shown below.

Time (minutes)	Cost ($)
14	1,510
80	6,178
84	5,912
118	9,184
149	8,855
192	11,023

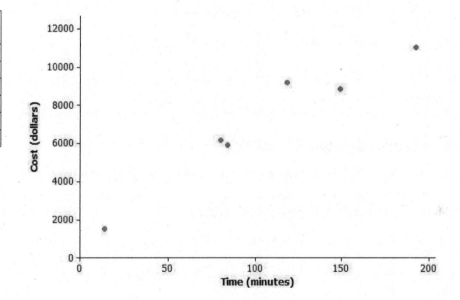

1. Calculate the equation of the least squares line relating cost to time. (Indicate slope to the nearest tenth and y-intercept to the nearest whole number.)

2. Draw the least squares line on the graph above. (Hint: Substitute $x = 30$ into your equation to find the predicted y-value. Plot the point (30, your answer) on the graph. Then substitute $x = 180$ into the equation, and plot the point. Join the two points with a straightedge.)

3. What does the least squares line predict for the cost of a surgery that lasts 118 min.? (Calculate the cost to the nearest cent.)

---

4. How much do you have to add to your answer to Problem 3 to get the actual cost of surgery for a surgery lasting 118 min.? (This is the residual.)

5. Show your answer to Problem 4 as a vertical line between the point for that person in the scatter plot and the least squares line.

6. Remember that the residual is the actual y-value minus the predicted y-value. Calculate the residual for the surgery that took 149 min. and cost $8,855.

7. Calculate the other residuals, and write all the residuals in the table below.

Time (minutes)	Cost ($)	Predicted Value ($)	Residual ($)
14	1,510		
80	6,178		
84	5,912		
118	9,184		
149	8,855		
192	11,023		

8. Suppose that a surgery took 100 min.

    a. What does the least squares line predict for the cost of this surgery?

    b. Would you be surprised if the actual cost of this surgery were $9,000? Why, or why not?

    c. Interpret the slope of the least squares line.

Lesson 15:   Interpreting Residuals from a Line

EUREKA
MATH™

Lesson 16: More on Modeling Relationships with a Line

Classwork

Example 1: Calculating Residuals

The curb weight of a car is the weight of the car without luggage or passengers. The table below shows the curb weights (in hundreds of pounds) and fuel efficiencies (in miles per gallon) of five compact cars.

Curb Weight (hundreds of pounds)	Fuel Efficiency (mpg)
25.33	43
26.94	38
27.79	30
30.12	34
32.47	30

Using a calculator, the least squares line for this data set was found to have the equation:

$$y = 78.62 - 1.5290x,$$

where x is the curb weight (in hundreds of pounds), and y is the predicted fuel efficiency (in miles per gallon).

The scatter plot of this data set is shown below, and the least squares line is shown on the graph.

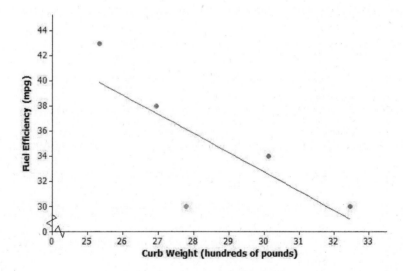

You will calculate the residuals for the five points in the scatter plot. Before calculating the residuals, look at the scatter plot.

EUREKA MATH™

Exercises 1–2

1. Will the residual for the car whose curb weight is 25.33 hundred pounds be positive or negative? Roughly what is the value of the residual for this point?

2. Will the residual for the car whose curb weight is 27.79 hundred pounds be positive or negative? Roughly what is the value of the residual for this point?

The residuals for both of these curb weights are calculated as follows:

Substitute $x = 25.33$ into the equation of the least squares line to find the predicted fuel efficiency. $$y = 78.62 - 1.5290(25.33)$$ $$= 39.9$$ Now calculate the residual. Residual = actual y-value − predicted y-value $$= 43 \text{ mpg} - 39.9 \text{ mpg}$$ $$= 3.1 \text{ mpg}$$	Substitute $x = 27.79$ into the equation of the least squares line to find the predicted fuel efficiency. $$y = 78.62 - 1.5290(27.79)$$ $$= 36.1$$ Now calculate the residual. residual = actual y-value − predicted y-value $$= 30 \text{ mpg} - 36.1 \text{ mpg}$$ $$= -6.1 \text{ mpg}$$

These two residuals have been written in the table below.

Curb Weight (hundreds of pounds)	Fuel Efficiency (mpg)	Residual (mpg)
25.33	43	3.1
26.94	38	
27.79	30	−6.1
30.12	34	
32.47	30	

EUREKA
MATH™

Exercises 3–4

Continue to think about the car weights and fuel efficiencies from Example 1.

3. Calculate the remaining three residuals, and write them in the table.

4. Suppose that a car has a curb weight of 31 hundred pounds.

    a. What does the least squares line predict for the fuel efficiency of this car?

    b. Would you be surprised if the actual fuel efficiency of this car was 29 miles per gallon? Explain your answer.

Example 2: Making a Residual Plot to Evaluate a Line

It is often useful to make a graph of the residuals, called a residual plot. You will make the residual plot for the compact car data set.

Plot the original x-variable (curb weight in this case) on the horizontal axis and the residuals on the vertical axis. For this example, you need to draw a horizontal axis that goes from 25 to 32 and a vertical axis with a scale that includes the values of the residuals that you calculated. Next, plot the point for the first car. The curb weight of the first car is 25.33 hundred pounds and the residual is 3.1 mpg. Plot the point $(25.33, 3.1)$.

The axes and this first point are shown below.

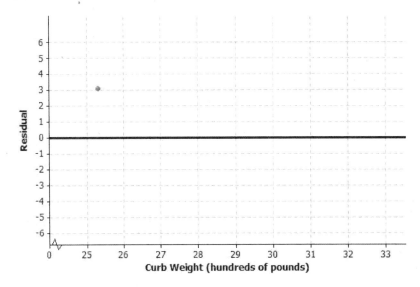

Exercise 5–6

5. Plot the other four residuals in the residual plot started in Example 3.

6. How does the pattern of the points in the residual plot relate to the pattern in the original scatter plot? Looking at the original scatter plot, could you have known what the pattern in the residual plot would be?

EUREKA
MATH™

> **Lesson Summary**
>
> - The predicted y-value is calculated using the equation of the least squares line.
> - The residual is calculated using
>
> $$\text{residual} = \text{actual } y\text{-value} - \text{predicted } y\text{-value}.$$
>
> - The sum of the residuals provides an idea of the degree of accuracy when using the least squares line to make predictions.
> - To make a residual plot, plot the x-values on the horizontal axis and the residuals on the vertical axis.

Problem Set

Four athletes on a track team are comparing their personal bests in the 100 meter and 200 meter events. A table of their best times is shown below.

Athlete	100 m time (seconds)	200 m time (seconds)
1	12.95	26.68
2	13.81	29.48
3	14.66	28.11
4	14.88	30.93

A scatter plot of these results (including the least squares line) is shown below.

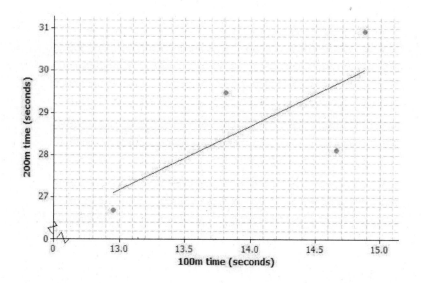

1. Use your calculator or computer to find the equation of the least squares line.

2. Use your equation to find the predicted 200-meter time for the runner whose 100-meter time is 12.95 seconds. What is the residual for this athlete?

3. Calculate the residuals for the other three athletes. Write all the residuals in the table given below.

Athlete	100 m time (seconds)	200 m time (seconds)	Residual (seconds)
1	12.95	26.68	
2	13.81	29.48	
3	14.66	28.11	
4	14.88	30.93	

4. Using the axes provided below, construct a residual plot for this data set.

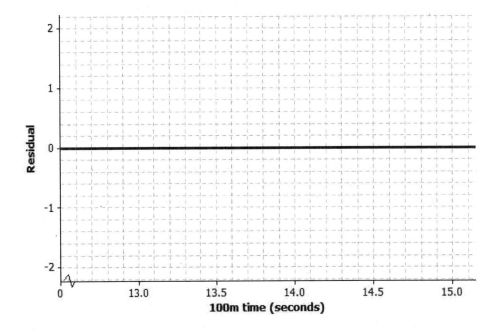

Lesson 16:   More on Modeling Relationships with a Line

Lesson 17: Analyzing Residuals

Classwork

Example 1: Predicting the Pattern in the Residual Plot

Suppose you are given a scatter plot and least squares line that looks like this:

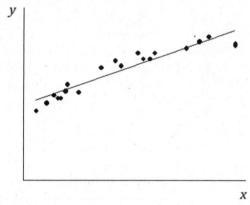

Describe what you think the residual plot would look like.

The residual plot has an arch shape like this:

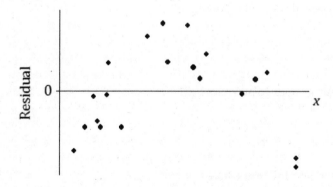

Why is looking at the pattern in the residual plot important?

Example 2: The Meaning of Residuals

Suppose that you have a scatter plot and that you have drawn the least squares line on your plot. Remember that the residual for a point in the scatter plot is the vertical distance of that point from the least squares line.

In the previous lesson, you looked at a scatter plot showing how fuel efficiency was related to curb weight for five compact cars. The scatter plot and least squares line are shown below.

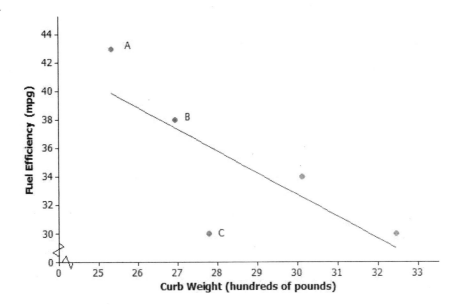

Consider the following questions:

- What kind of residual does Point A have?

- What kind of residual does Point B have?

- What kind of residual does Point C have?

EUREKA
MATH

You also looked at the residual plot for this data set:

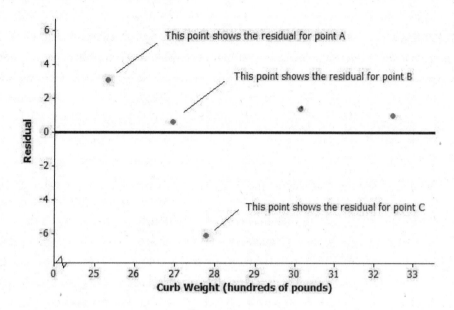

Your teacher will now show how to use a graphing calculator or graphing program to construct a scatter plot and a residual plot. Consider the following example.

Example 3: Using a Graphing Calculator to Construct a Residual Plot

In an earlier lesson, you looked at a data set giving the shoe lengths and heights of 12 adult women. This data set is shown in the table below.

x (Shoe Length)	y (Height)
inches	inches
8.9	61
9.6	61
9.8	66
10.0	64
10.2	64
10.4	65
10.6	65
10.6	67
10.5	66
10.8	67
11.0	67
11.8	70

Use a calculator to construct the scatter plot (with least squares line) and the residual plot for this data set.

Lesson Summary

- After fitting a line, the residual plot can be constructed using a graphing calculator.
- A pattern in the residual plot indicates that the relationship in the original data set is not linear.

Problem Set

Consider again a data set giving the shoe lengths and heights of 10 adult men. This data set is shown in the table below.

x (Shoe Length)	y (Height)
inches	inches
12.6	74
11.8	65
12.2	71
11.6	67
12.2	69
11.4	68
12.8	70
12.2	69
12.6	72
11.8	71

1. Use your calculator or graphing program to construct the scatter plot of this data set. Include the least squares line on your graph. Explain what the slope of the least squares line indicates about shoe length and height.

2. Use your calculator to construct the residual plot for this data set.

3. Make a sketch of the residual plot on the axes given below. Does the scatter of points in the residual plot indicate a linear relationship in the original data set? Explain your answer.

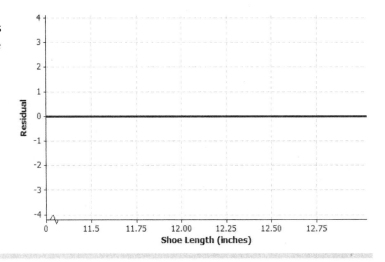

Lesson 17:   Analyzing Residuals

Lesson 18: Analyzing Residuals

Classwork

The previous lesson shows that when data is fitted to a line, a scatter plot with a curved pattern produces a residual plot that shows a clear pattern. You also saw that when a line is fit, a scatter plot where the points show a straight-line pattern results in a residual plot where the points are randomly scattered.

Example 1: The Relevance of the Pattern in the Residual Plot

Our previous findings are summarized in the plots below:

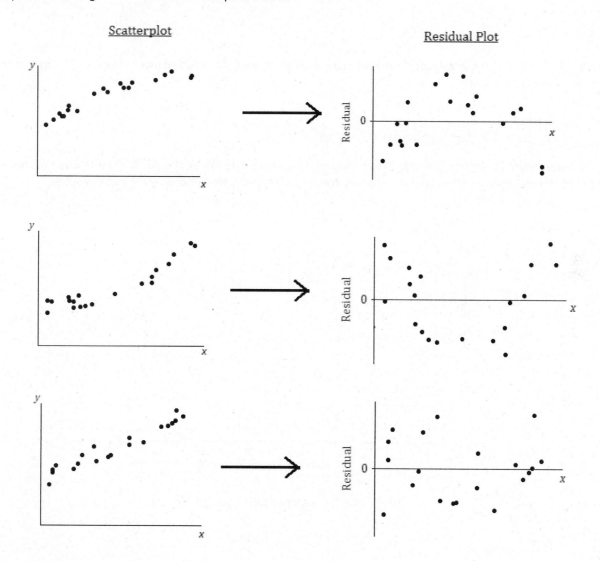

What does it mean when there is a curved pattern in the residual plot?

What does it mean when the points in the residual plot appear to be scattered at random with no visible pattern?

Why not just look at the scatter plot of the original data set? Why was the residual plot necessary? The next example answers these questions.

Example 2: Why Do You Need the Residual Plot?

The temperature (in degrees Fahrenheit) was measured at various altitudes (in thousands of feet) above Los Angeles. The scatter plot (below) seems to show a linear (straight-line) relationship between these two quantities.

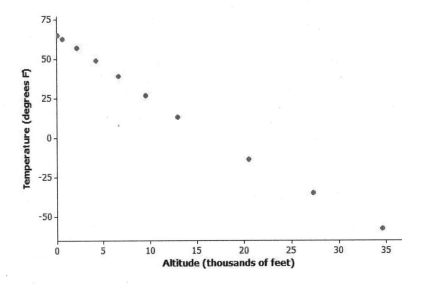

Data source: *Core Math Tools,* http://nctm.org

EUREKA
MATH™

However, look at the residual plot:

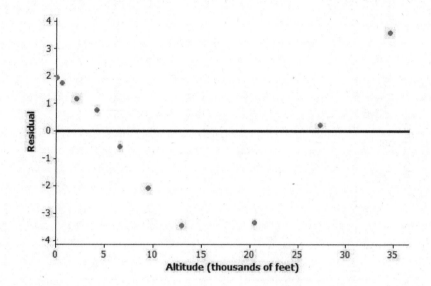

There is a clear curve in the residual plot. So what appeared to be a linear relationship in the original scatter plot was, in fact, a nonlinear relationship.

How did this residual plot result from the original scatter plot?

Exercises 1–3: Volume and Temperature

Water expands as it heats. Researchers measured the volume (in milliliters) of water at various temperatures. The results are shown below.

Temperature (°C)	Volume (ml)
20	100.125
21	100.145
22	100.170
23	100.191
24	100.215
25	100.239
26	100.266
27	100.290
28	100.319
29	100.345
30	100.374

1. Using a graphing calculator, construct the scatter plot of this data set. Include the least squares line on your graph. Make a sketch of the scatter plot including the least squares line on the axes below.

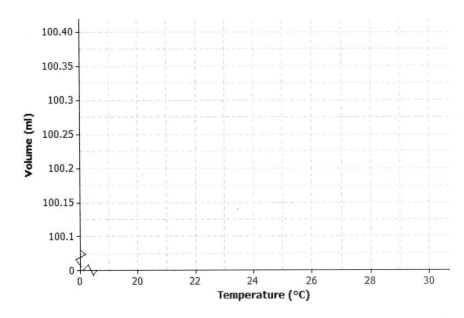

2. Using the calculator, construct a residual plot for this data set. Make a sketch of the residual plot on the axes given below.

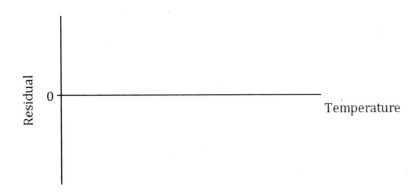

3. Do you see a clear curve in the residual plot? What does this say about the original data set?

Lesson 18:   Analyzing Residuals

EUREKA
MATH™

Lesson Summary

- After fitting a line, the residual plot can be constructed using a graphing calculator.
- A curve or pattern in the residual plot indicates a nonlinear relationship in the original data set.
- A random scatter of points in the residual plot indicates a linear relationship in the original data set.

Problem Set

1. For each of the following residual plots, what conclusion would you reach about the relationship between the variables in the original data set? Indicate whether the values would be better represented by a linear or a nonlinear relationship.

    a.

    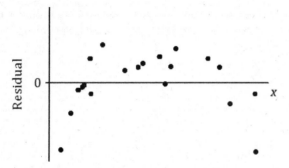

    b.

    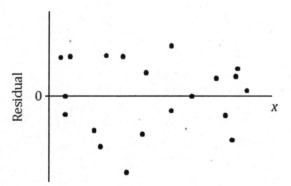

    c.

    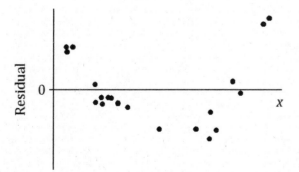

2. Suppose that after fitting a line, a data set produces the residual plot shown below.

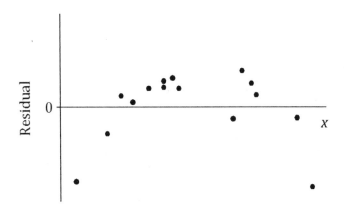

An incomplete scatter plot of the original data set is shown below. The least squares line is shown, but the points in the scatter plot have been erased. Estimate the locations of the original points, and create an approximation of the scatter plot below.

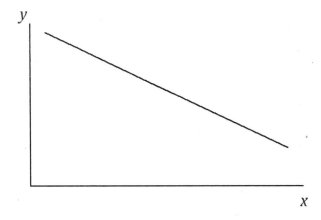

Lesson 18:   Analyzing Residuals

EUREKA
MATH™

Lesson 19: Interpreting Correlation

Classwork

Example 1: Positive and Negative Linear Relationships

Linear relationships can be described as either positive or negative. Below are two scatter plots that display a linear relationship between two numerical variables x and y.

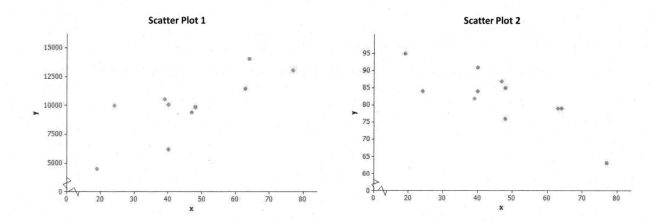

Exercises 1–4

1. The relationship displayed in Scatter Plot 1 is a positive linear relationship. Does the value of the y-variable tend to increase or decrease as the value of x increases? If you were to describe this relationship using a line, would the line have a positive or negative slope?

2. The relationship displayed in Scatter Plot 2 is a negative linear relationship. As the value of one of the variables increases, what happens to the value of the other variable? If you were to describe this relationship using a line, would the line have a positive or negative slope?

3. What does it mean to say that there is a positive linear relationship between two variables?

4. What does it mean to say that there is a negative linear relationship between two variables?

Example 2: Some Linear Relationships Are Stronger than Others

Below are two scatter plots that show a linear relationship between two numerical variables x and y.

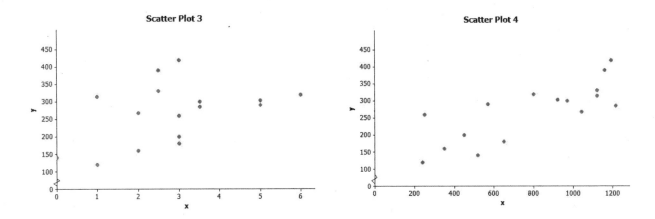

Exercises 5–9

5. Is the linear relationship in Scatter Plot 3 positive or negative?

6. Is the linear relationship in Scatter Plot 4 positive or negative?

EUREKA
MATH™

It is also common to describe the strength of a linear relationship. We would say that the linear relationship in Scatter Plot 3 is weaker than the linear relationship in Scatter Plot 4.

7. Why do you think the linear relationship in Scatter Plot 3 is considered weaker than the linear relationship in Scatter Plot 4?

8. What do you think a scatter plot with the strongest possible linear relationship might look like if it is a positive relationship? Draw a scatter plot with five points that illustrates this.

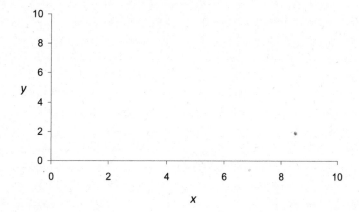

9. How would a scatter plot that shows the strongest possible linear relationship that is negative look different from the scatter plot that you drew in the previous question?

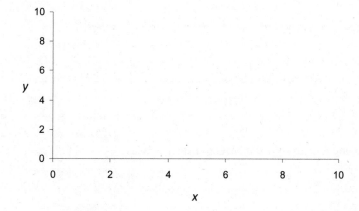

Exercises 10–12: Strength of Linear Relationships

10. Consider the three scatter plots below. Place them in order from the one that shows the strongest linear relationship to the one that shows the weakest linear relationship.

Strongest	⟶	Weakest

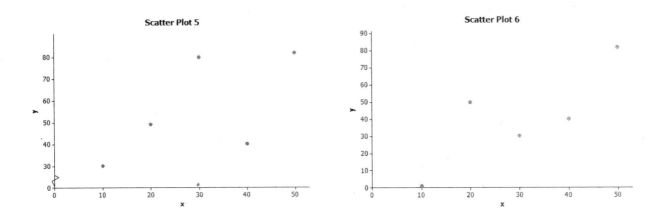

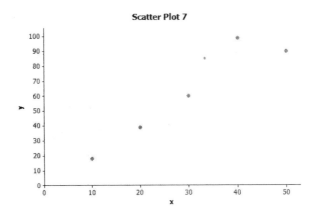

11. Explain your reasoning for choosing the order in Exercise 10.

Lesson 19:   Interpreting Correlation

EUREKA
MATH™

12. Which of the following two scatter plots shows the stronger linear relationship? (Think carefully about this one!)

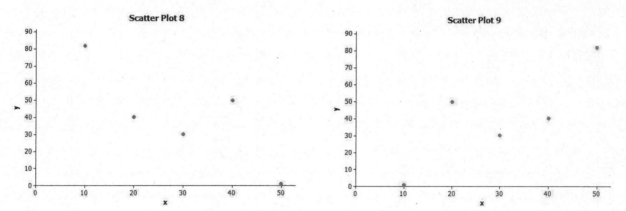

Example 3: The Correlation Coefficient

The *correlation coefficient* is a number between -1 and $+1$ (including -1 and $+1$) that measures the strength and direction of a linear relationship. The correlation coefficient is denoted by the letter r.

Several scatter plots are shown below. The value of the correlation coefficient for the data displayed in each plot is also given.

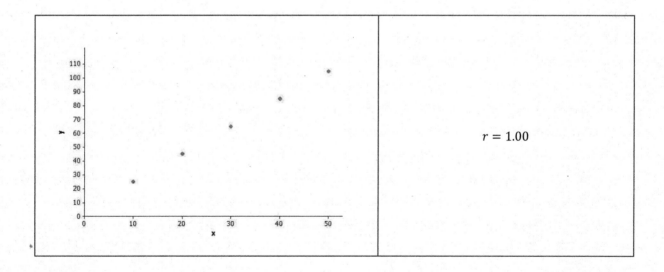

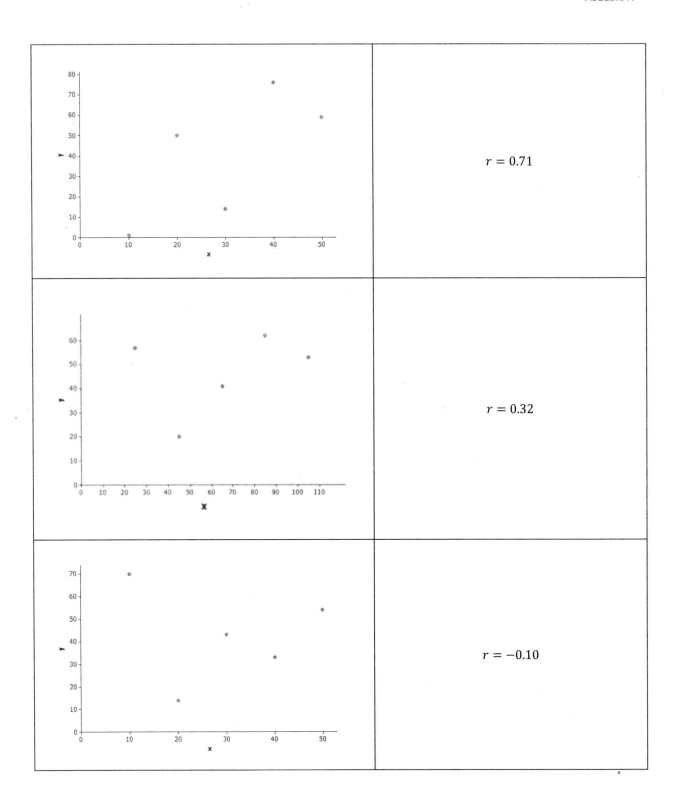

Lesson 19: Interpreting Correlation

EUREKA
MATH™

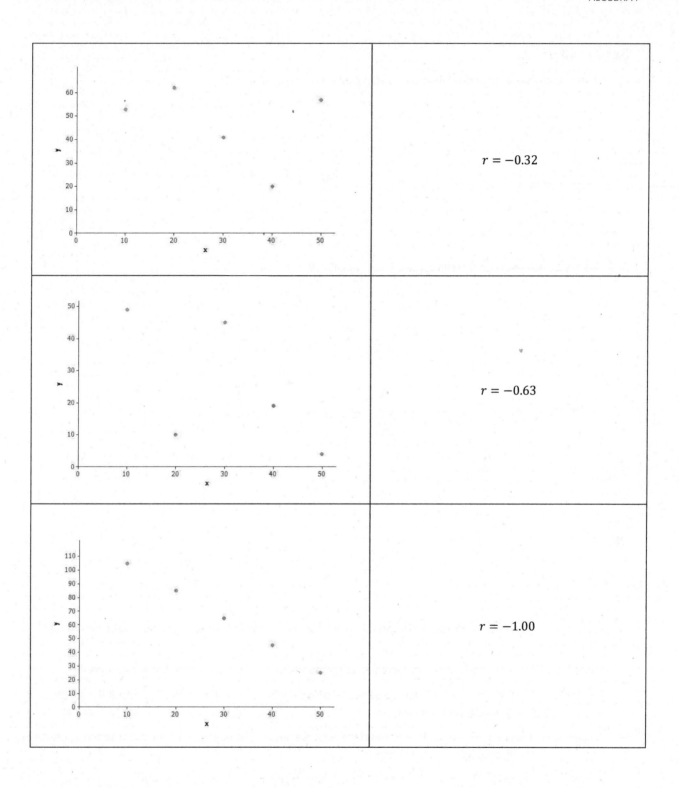

$r = -0.32$

$r = -0.63$

$r = -1.00$

Exercises 13–15

13. When is the value of the correlation coefficient positive?

14. When is the value of the correlation coefficient negative?

15. Is the linear relationship stronger when the correlation coefficient is closer to 0 or to 1 (or −1)?

Looking at the scatter plots in Example 4, you should have discovered the following properties of the correlation coefficient:

Property 1:   The sign of r (positive or negative) corresponds to the direction of the linear relationship.

Property 2:   A value of $r = +1$ indicates a perfect positive linear relationship, with all points in the scatter plot falling exactly on a straight line.

Property 3:   A value of $r = -1$ indicates a perfect negative linear relationship, with all points in the scatter plot falling exactly on a straight line.

Property 4:   The closer the value of r is to $+1$ or -1, the stronger the linear relationship.

Example 4: Calculating the Value of the Correlation Coefficient

There is an equation that can be used to calculate the value of the correlation coefficient given data on two numerical variables. Using this formula requires a lot of tedious calculations that are discussed in later grades. Fortunately, a graphing calculator can be used to find the value of the correlation coefficient once you have entered the data.

Your teacher will show you how to enter data and how to use a graphing calculator to obtain the value of the correlation coefficient.

Here is the data from a previous lesson on shoe length in inches and height in inches for 10 men.

x (Shoe Length) inches	y (Height) inches
12.6	74
11.8	65
12.2	71
11.6	67
12.2	69
11.4	68
12.8	70
12.2	69
12.6	72
11.8	71

Exercises 16–17

16. Enter the shoe length and height data in your calculator. Find the value of the correlation coefficient between shoe length and height. Round to the nearest tenth.

The table below shows how you can informally interpret the value of a correlation coefficient.

If the value of the correlation coefficient is between …	You can say that …
$r = 1.0$	There is a perfect positive linear relationship.
$0.7 \leq r < 1.0$	There is a strong positive linear relationship.
$0.3 \leq r < 0.7$	There is a moderate positive linear relationship.
$0 < r < 0.3$	There is a weak positive linear relationship.
$r = 0$	There is no linear relationship.
$-0.3 < r < 0$	There is a weak negative linear relationship.
$-0.7 < r \leq -0.3$	There is a moderate negative linear relationship.
$-1.0 < r \leq -0.7$	There is a strong negative linear relationship.
$r = -1.0$	There is a perfect negative linear relationship.

17. Interpret the value of the correlation coefficient between shoe length and height for the data given above.

Exercises 18–24: Practice Calculating and Interpreting Correlation Coefficients

Consumer Reports published a study of fast-food items. The table and scatter plot below display the fat content (in grams) and number of calories per serving for 16 fast-food items.

Fat (g)	Calories (kcal)
2	268
5	303
3	260
3.5	300
1	315
2	160
3	200
6	320
3	420
5	290
3.5	285
2.5	390
0	140
2.5	330
1	120
3	180

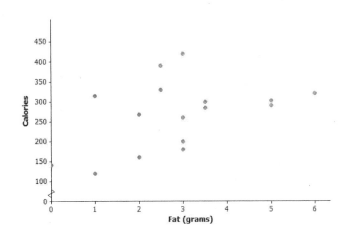

Data Source: *Consumer Reports*

18. Based on the scatter plot, do you think that the value of the correlation coefficient between fat content and calories per serving will be positive or negative? Explain why you made this choice.

Lesson 19:   Interpreting Correlation

19. Based on the scatter plot, estimate the value of the correlation coefficient between fat content and calories.

20. Calculate the value of the correlation coefficient between fat content and calories per serving. Round to the nearest hundredth. Interpret this value.

The *Consumer Reports* study also collected data on sodium content (in mg) and number of calories per serving for the same 16 fast food items. The data is represented in the table and scatter plot below.

Sodium (mg)	Calories (kcal)
1,042	268
921	303
250	260
970	300
1,120	315
350	160
450	200
800	320
1,190	420
570	290
1,215	285
1,160	390
520	140
1,120	330
240	120
650	180

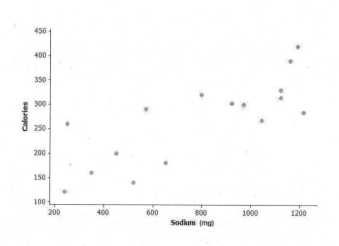

21. Based on the scatter plot, do you think that the value of the correlation coefficient between sodium content and calories per serving will be positive or negative? Explain why you made this choice.

22. Based on the scatter plot, estimate the value of the correlation coefficient between sodium content and calories per serving.

23. Calculate the value of the correlation coefficient between sodium content and calories per serving. Round to the nearest hundredth. Interpret this value.

24. For these 16 fast-food items, is the linear relationship between fat content and number of calories stronger or weaker than the linear relationship between sodium content and number of calories? Does this surprise you? Explain why or why not.

Example 5: Correlation Does Not Mean There is a Cause-and-Effect Relationship Between Variables

It is sometimes tempting to conclude that if there is a strong linear relationship between two variables that one variable is causing the value of the other variable to increase or decrease. But you should avoid making this mistake. When there is a strong linear relationship, it means that the two variables tend to vary together in a predictable way, which might be due to something other than a cause-and-effect relationship.

For example, the value of the correlation coefficient between sodium content and number of calories for the fast food items in the previous example was $r = 0.79$, indicating a strong positive relationship. This means that the items with higher sodium content tend to have a higher number of calories. But the high number of calories is not caused by the high sodium content. In fact, sodium does not have any calories. What may be happening is that food items with high sodium content also may be the items that are high in sugar or fat, and this is the reason for the higher number of calories in these items.

Similarly, there is a strong positive correlation between shoe size and reading ability in children. But it would be silly to think that having big feet causes children to read better. It just means that the two variables vary together in a predictable way. Can you think of a reason that might explain why children with larger feet also tend to score higher on reading tests?

Lesson Summary

- Linear relationships are often described in terms of strength and direction.
- The correlation coefficient is a measure of the strength and direction of a linear relationship.
- The closer the value of the correlation coefficient is to $+1$ or -1, the stronger the linear relationship.
- Just because there is a strong correlation between the two variables does not mean there is a cause-and-effect relationship.

Problem Set

1. Which of the three scatter plots below shows the strongest linear relationship? Which shows the weakest linear relationship?

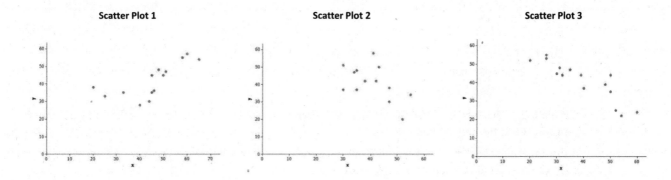

2. *Consumer Reports* published data on the price (in dollars) and quality rating (on a scale of 0 to 100) for 10 different brands of men's athletic shoes.

Price ($)	Quality Rating
65	71
45	70
45	62
80	59
110	58
110	57
30	56
80	52
110	51
70	51

a. Construct a scatter plot of these data using the grid provided.

b. Calculate the value of the correlation coefficient between price and quality rating, and interpret this value. Round to the nearest hundredth.

c. Does it surprise you that the value of the correlation coefficient is negative? Explain why or why not.

d. Is it reasonable to conclude that higher-priced shoes are higher quality? Explain.

e. The correlation between price and quality rating is negative. Is it reasonable to conclude that increasing the price causes a decrease in quality rating? Explain.

3. *The Princeton Review* publishes information about colleges and universities. The data below are for six public 4-year colleges in New York. Graduation rate is the percentage of students who graduate within six years. Student-to-faculty ratio is the number of students per full-time faculty member.

School	Number of Full-Time Students	Student-to-Faculty Ratio	Graduation Rate
CUNY Bernard M. Baruch College	11,477	17	63
CUNY Brooklyn College	9,876	15.3	48
CUNY City College	10,047	13.1	40
SUNY at Albany	14,013	19.5	64
SUNY at Binghamton	13,031	20	77
SUNY College at Buffalo	9,398	14.1	47

a. Calculate the value of the correlation coefficient between the number of full-time students and graduation rate. Round to the nearest hundredth.

b. Is the linear relationship between graduation rate and number of full-time students weak, moderate, or strong? On what did you base your decision?

c. Is the following statement true or false? Based on the value of the correlation coefficient, it is reasonable to conclude that having a larger number of students at a school is the cause of a higher graduation rate.

d. Calculate the value of the correlation coefficient between the student-to-faculty ratio and the graduation rate. Round to the nearest hundredth.

e. Which linear relationship is stronger: graduation rate and number of full-time students or graduation rate and student-to-faculty ratio? Justify your choice.

Lesson 20: Analyzing Data Collected on Two Variables

Classwork

Lessons 12–19 included several data sets that were used to learn about how two numerical variables might be related. Recall the data on elevation above sea level and the number of clear days per year for 14 cities in the United States. Could a city's elevation above sea level by used to predict the number of clear days per year a city experiences? After observing a scatter plot of the data, a linear model (or the least squares linear model obtained from a calculator or computer software) provided a reasonable description of the relationship between these two variables. This linear model was evaluated by considering how close the data points were to the corresponding graph of the line. The equation of the linear model was used to answer the statistical question about elevation and the number of clear days.

Several data sets were also provided to illustrate other possible models, specifically quadratic and exponential models. Finding a model that describes the relationship between two variables allows you to answer statistical questions about how the two numerical variables vary. For example, the following statistical question was posed in Lesson 13 regarding latitude and the mean number of flycatcher chicks in a nest: What latitude is best for hatching flycatcher chicks? A quadratic model of the latitude and the mean number of flycatcher chicks in a nest was used to answer this statistical question.

In Lessons 12–19, you worked with several data sets and models used to answer statistical questions. Select one of the data sets presented in Lessons 12–19, and develop a poster that summarizes how a statistical question is answered that involves two numerical variables. Your poster should include the following: a brief summary of the data, the statistical question asking the relationship between two numerical variables, a scatter plot of the data, and a brief summary to indicate how well the data fit a specific model.

After you identify one of the problems to summarize with a poster, consider the following questions to plan your poster:

1. What two variables were involved in this problem?

2. What was the statistical question? Remember, a statistical question involves data. The question also anticipates that the data will vary.

3. What model was used to describe the relationship between the two variables?

4. Based on the scatter plot of the data, was the model a good one?

5. Was the residual plot used to evaluate the model? What did the residual plot indicate about the model?

6. How would you use the model to predict values not included in the data set?

7. Does the model answer the statistical question posed?

Examples of posters involving two numerical variables can be found at the website of the American Statistical Association (www.amstat.org/education/posterprojects/index.cfm).

This page intentionally left blank

Eureka Math
Algebra I
Module 3

Special thanks go to the Gordan A. Cain Center and to the Department of Mathematics at Louisiana State University for their support in the development of *Eureka Math*.

For a free *Eureka Math* Teacher Resource Pack, Parent Tip Sheets, and more please visit www.Eureka.tools

Published by Great Minds

Copyright © 2015 Great Minds. All rights reserved. No part of this work may be reproduced or used in any form or by any means — graphic, electronic, or mechanical, including photocopying or information storage and retrieval systems — without written permission from the copyright holder. "Great Minds" and "Eureka Math" are registered trademarks of Great Minds.

Printed in the U.S.A.

This book may be purchased from the publisher at eureka-math.org

10 9 8 7 6 5 4 3 2 1

ISBN 978-1-63255-324-9

Lesson 1: Integer Sequences—Should You Believe in Patterns?

Classwork

Opening Exercise

Mrs. Rosenblatt gave her students what she thought was a very simple task:

What is the next number in the sequence 2, 4, 6, 8, …?

Cody: I am thinking of a plus 2 pattern, so it continues 10, 12, 14, 16, ….

Ali: I am thinking of a repeating pattern, so it continues 2, 4, 6, 8, 2, 4, 6, 8, ….

Suri: I am thinking of the units digits in the multiples of two, so it continues 2, 4, 6, 8, 0, 2, 4, 6, 8, ….

    a.   Are each of these valid responses?

    b.   What is the hundredth number in the sequence in Cody's scenario? Ali's? Suri's?

    c.   What is an expression in terms of n for the nth number in the sequence in Cody's scenario?

Example 1

Jerry has thought of a pattern that shows powers of two. Here are the first six numbers of Jerry's sequence:

      1, 2, 4, 8, 16, 32, ….

Write an expression for the nth number of Jerry's sequence.

Example 2

Consider the sequence that follows a plus 3 pattern: $4, 7, 10, 13, 16, \dots$.

    a.   Write a formula for the sequence using both the a_n notation and the $f(n)$ notation.

    b.   Does the formula $f(n) = 3(n-1) + 4$ generate the same sequence? Why might some people prefer this formula?

    c.   Graph the terms of the sequence as ordered pairs $\big(n, f(n)\big)$ on the coordinate plane. What do you notice about the graph?

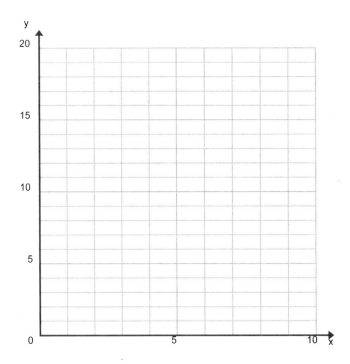

      Lesson 1:   Integer Sequences—Should You Believe in Patterns?

EUREKA
MATH™

Exercises

1. Refer back to the sequence from the Opening Exercise. When Mrs. Rosenblatt was asked for the next number in the sequence 2, 4, 6, 8, …, she said "17." The class responded, "17?"

   Yes, using the formula, $f(n) = \frac{7}{24}(n-1)^4 - \frac{7}{4}(n-1)^3 + \frac{77}{24}(n-1)^2 + \frac{1}{4}(n-1) + 2$.

   a. Does her formula actually produce the numbers 2, 4, 6, and 8?

   b. What is the 100th term in Mrs. Rosenblatt's sequence?

2. Consider a sequence that follows a minus 5 pattern: 30, 25, 20, 15, ….

   a. Write a formula for the nth term of the sequence. Be sure to specify what value of n your formula starts with.

   b. Using the formula, find the 20th term of the sequence.

   c. Graph the terms of the sequence as ordered pairs $\big(n, f(n)\big)$ on a coordinate plane.

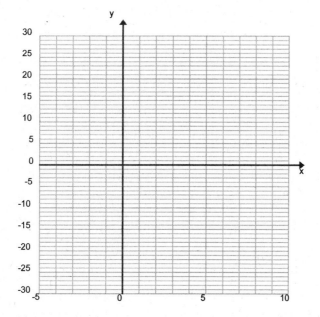

3. Consider a sequence that follows a times 5 pattern: 1, 5, 25, 125, ….

   a. Write a formula for the nth term of the sequence. Be sure to specify what value of n your formula starts with.

   b. Using the formula, find the 10th term of the sequence.

   c. Graph the terms of the sequence as ordered pairs $\big(n, f(n)\big)$ on a coordinate plane.

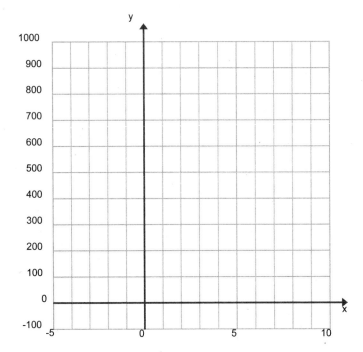

EUREKA
MATH™

This work is derived from Eureka Math ™ and licensed by Great Minds. ©2015 Great Minds. eureka-math.org
ALG1-M3-SE-B1-1.3.0-05.2015

4. Consider the sequence formed by the square numbers:

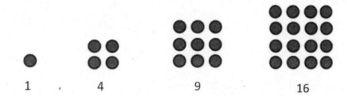

1   .   4   9   16

a. Write a formula for the nth term of the sequence. Be sure to specify what value of n your formula starts with.

b. Using the formula, find the 50th term of the sequence.

c. Graph the terms of the sequence as ordered pairs $\big(n, f(n)\big)$ on a coordinate plane.

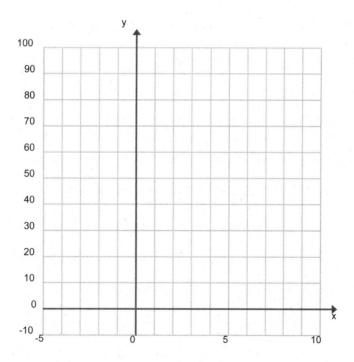

5. A standard letter-sized piece of paper has a length and width of 8.5 inches by 11 inches.

   a. Find the area of one piece of paper.

   b. If the paper were folded completely in half, what would be the area of the resulting rectangle?

   c. Write a formula for a sequence to determine the area of the paper after n folds.

   d. What would the area be after 7 folds?

S.6   **Lesson 1:**   Integer Sequences—Should You Believe in Patterns?

**EUREKA
MATH**™

Lesson Summary

Think of a sequence as an ordered list of elements. Give an explicit formula to define the pattern of the sequence. Unless specified otherwise, find the first term by substituting 1 into the formula.

Problem Set

1. Consider a sequence generated by the formula $f(n) = 6n - 4$ starting with $n = 1$. Generate the terms $f(1)$, $f(2)$, $f(3)$, $f(4)$, and $f(5)$.

2. Consider a sequence given by the formula $f(n) = \dfrac{1}{3^{n-1}}$ starting with $n = 1$. Generate the first 5 terms of the sequence.

3. Consider a sequence given by the formula $f(n) = (-1)^n \times 3$ starting with $n = 1$. Generate the first 5 terms of the sequence.

4. Here is the classic puzzle that shows that patterns need not hold true. What are the numbers counting?

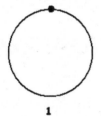
1

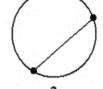
2

4

8

16

??

    a. Based on the sequence of numbers, predict the next number.
    b. Write a formula based on the perceived pattern.
    c. Find the next number in the sequence by actually counting.
    d. Based on your answer from part (c), is your model from part (b) effective for this puzzle?

For each of the sequences in Problems 5–8:

    a.   Write a formula for the nth term of the sequence. Be sure to specify what value of n your formula starts with.

    b.   Using the formula, find the 15th term of the sequence.

    c.   Graph the terms of the sequence as ordered pairs $\left(n, f(n)\right)$ on a coordinate plane.

5.   The sequence follows a plus 2 pattern: $3, 5, 7, 9, \ldots.$

6.   The sequence follows a times 4 pattern: $1, 4, 16, 64, \ldots.$

7.   The sequence follows a times -1 pattern: $6, -6, 6, -6, \ldots.$

8.   The sequence follows a minus 3 pattern: $12, 9, 6, 3, \ldots.$

     Lesson 1:   Integer Sequences—Should You Believe in Patterns?

EUREKA MATH

Lesson 2: Recursive Formulas for Sequences

Classwork

Example 1

Consider Akelia's sequence 5, 8, 11, 14, 17, ...

    a.   If you believed in patterns, what might you say is the next number in the sequence?

    b.   Write a formula for Akelia's sequence.

    c.   Explain how each part of the formula relates to the sequence.

    d.   Explain Johnny's formula.

Exercises 1–2

1.   Akelia, in a playful mood, asked Johnny: "What would happen if we change the '+' sign in your formula to a '−' sign? To a '×' sign? To a '÷' sign?"

    a.   What sequence does $A(n + 1) = A(n) - 3$ for $n \geq 1$ and $A(1) = 5$ generate?

    b.   What sequence does $A(n + 1) = A(n) \cdot 3$ for $n \geq 1$ and $A(1) = 5$ generate?

    c.   What sequence does $A(n + 1) = A(n) \div 3$ for $n \geq 1$ and $A(1) = 5$ generate?

2. Ben made up a recursive formula and used it to generate a sequence. He used $B(n)$ to stand for the nth term of his recursive sequence.

a. What does $B(3)$ mean?

b. What does $B(m)$ mean?

c. If $B(n+1) = 33$ and $B(n) = 28$, write a possible recursive formula involving $B(n+1)$ and $B(n)$ that would generate 28 and 33 in the sequence.

d. What does $2B(7) + 6$ mean?

e. What does $B(n) + B(m)$ mean?

f. Would it necessarily be the same as $B(n+m)$?

g. What does $B(17) - B(16)$ mean?

Lesson 2:   Recursive Formulas for Sequences

EUREKA MATH™

Example 2

Consider a sequence given by the formula $a_n = a_{n-1} - 5$, where $a_1 = 12$ and $n \geq 2$.

    a.   List the first five terms of the sequence.

    b.   Write an explicit formula.

    c.   Find a_6 and a_{100} of the sequence.

Exercises 3–6

3.   One of the most famous sequences is the Fibonacci sequence:

    1, 1, 2, 3, 5, 8, 13, 21, 34, ….

    $f(n + 1) = f(n) + f(n - 1)$, where $f(1) = 1$, $f(2) = 1$, and $n \geq 2$.

    How is each term of the sequence generated?

4.   Each sequence below gives an explicit formula. Write the first five terms of each sequence. Then, write a recursive formula for the sequence.

    a. $a_n = 2n + 10$ for $n \geq 1$

b. $a_n = \left(\frac{1}{2}\right)^{n-1}$ for $n \geq 1$

5. For each sequence, write *either* an explicit or a recursive formula.

a. $1, -1, 1, -1, 1, -1, \ldots$

b. $\dfrac{1}{2}, \dfrac{2}{3}, \dfrac{3}{4}, \dfrac{4}{5}, \ldots$

6. Lou opens a bank account. The deal he makes with his mother is that if he doubles the amount that was in the account at the beginning of each month by the end of the month, she will add an additional $5 to the account at the end of the month.

a.   Let $A(n)$ represent the amount in the account at the beginning of the nth month. Assume that he does, in fact, double the amount every month. Write a recursive formula for the amount of money in his account at the beginning of the $(n + 1)^{\text{th}}$ month.

b.   What is the least amount he could start with in order to have $300 by the beginning of the third month?

EUREKA
MATH™

> **Lesson Summary**
>
> **RECURSIVE SEQUENCE:** An example of a *recursive sequence* is a sequence that (1) is defined by specifying the values of one or more initial terms and (2) has the property that the remaining terms satisfy a recursive formula that describes the value of a term based upon an expression in numbers, previous terms, or the index of the term.
>
> An explicit formula specifies the nth term of a sequence as an expression in n.
>
> A recursive formula specifies the nth term of a sequence as an expression in the previous term (or previous couple of terms).

Problem Set

For Problems 1–4, list the first five terms of each sequence.

1. $a_{n+1} = a_n + 6$, where $a_1 = 11$ for $n \geq 1$

2. $a_n = a_{n-1} \div 2$, where $a_1 = 50$ for $n \geq 2$

3. $f(n + 1) = -2f(n) + 8$ and $f(1) = 1$ for $n \geq 1$

4. $f(n) = f(n - 1) + n$ and $f(1) = 4$ for $n \geq 2$

For Problems 5–10, write a recursive formula for each sequence given or described below.

5. It follows a plus one pattern: 8, 9, 10, 11, 12, ….

6. It follows a times 10 pattern: 4, 40, 400, 4000, ….

7. It has an explicit formula of $f(n) = -3n + 2$ for $n \geq 1$.

8. It has an explicit formula of $f(n) = -1(12)^{n-1}$ for $n \geq 1$.

9. Doug accepts a job where his starting salary is \$30,000 per year, and each year he receives a raise of \$3,000.

10. A bacteria culture has an initial population of 10 bacteria, and each hour the population triples in size.

This page intentionally left blank

Lesson 3: Arithmetic and Geometric Sequences

Classwork

Exercise 2

Think of a real-world example of an arithmetic or a geometric sequence. Describe it, and write its formula.

Exercise 3

If we fold a rectangular piece of paper in half multiple times and count the number of rectangles created, what type of sequence are we creating? Can you write the formula?

> **Lesson Summary**
>
> Two types of sequences were studied:
>
> **ARITHMETIC SEQUENCE**: A sequence is called *arithmetic* if there is a real number d such that each term in the sequence is the sum of the previous term and d.
>
> **GEOMETRIC SEQUENCE**: A sequence is called *geometric* if there is a real number r such that each term in the sequence is a product of the previous term and r.

Problem Set

For Problems 1–4, list the first five terms of each sequence, and identify them as arithmetic or geometric.

1. $A(n + 1) = A(n) + 4$ for $n \geq 1$ and $A(1) = -2$

2. $A(n + 1) = \frac{1}{4} \cdot A(n)$ for $n \geq 1$ and $A(1) = 8$

3. $A(n + 1) = A(n) - 19$ for $n \geq 1$ and $A(1) = -6$

4. $A(n + 1) = \frac{2}{3} A(n)$ for $n \geq 1$ and $A(1) = 6$

For Problems 5–8, identify the sequence as arithmetic or geometric, and write a recursive formula for the sequence. Be sure to identify your starting value.

5. 14, 21, 28, 35, …

6. 4, 40, 400, 4000, …

7. $49, 7, 1, \frac{1}{7}, \frac{1}{49}, \dots$

8. −101, −91, −81, −71, …

9. The local football team won the championship several years ago, and since then, ticket prices have been increasing $20 per year. The year they won the championship, tickets were $50. Write a recursive formula for a sequence that models ticket prices. Is the sequence arithmetic or geometric?

EUREKA
MATH™

10. A radioactive substance decreases in the amount of grams by one-third each year. If the starting amount of the substance in a rock is 1,452 g, write a recursive formula for a sequence that models the amount of the substance left after the end of each year. Is the sequence arithmetic or geometric?

11. Find an explicit form $f(n)$ for each of the following arithmetic sequences (assume a is some real number and x is some real number).

    a. $-34, -22, -10, 2, \ldots$

    b. $\dfrac{1}{5}, \dfrac{1}{10}, 0, -\dfrac{1}{10}, \ldots$

    c. $x + 4, x + 8, x + 12, x + 16, \ldots$

    d. $a, 2a + 1, 3a + 2, 4a + 3, \ldots$

12. Consider the arithmetic sequence $13, 24, 35, \ldots$.

    a. Find an explicit form for the sequence in terms of n.

    b. Find the 40th term.

    c. If the nth term is 299, find the value of n.

13. If $-2, a, b, c, 14$ forms an arithmetic sequence, find the values of $a, b,$ and c.

14. $3 + x, 9 + 3x, 13 + 4x, \ldots$ is an arithmetic sequence for some real number x.

    a. Find the value of x.

    b. Find the 10th term of the sequence.

15. Find an explicit form $f(n)$ of the arithmetic sequence where the 2nd term is 25 and the sum of the 3rd term and 4th term is 86.

16. Challenge: In the right triangle figure below, the lengths of the sides a cm, b cm, and c cm of the right triangle form a finite arithmetic sequence. If the perimeter of the triangle is 18 cm, find the values of $a, b,$ and c.

17. Find the common ratio and an explicit form in each of the following geometric sequences.

    a. $4, 12, 36, 108, \ldots$

    b. $162, 108, 72, 48, \ldots$

    c. $\dfrac{4}{3}, \dfrac{2}{3}, \dfrac{1}{3}, \dfrac{1}{6}, \ldots$

    d. $xz, x^2z^3, x^3z^5, x^4z^7, \ldots$

18. The first term in a geometric sequence is 54, and the 5th term is $\frac{2}{3}$. Find an explicit form for the geometric sequence.

19. If $2, a, b, -54$ forms a geometric sequence, find the values of a and b.

20. Find the explicit form $f(n)$ of a geometric sequence if $f(3) - f(1) = 48$ and $\dfrac{f(3)}{f(1)} = 9$.

Lesson 4: Why Do Banks Pay YOU to Provide Their Services?

Classwork

Example 1

Kyra has been babysitting since sixth grade. She has saved $1,000 and wants to open an account at the bank so that she earns interest on her savings. Simple Bank pays simple interest at a rate of 10%. How much money will Kyra have after 1 year? After 2 years, if she does not add money to her account? After 5 years?

Raoul needs $200 to start a snow cone stand for this hot summer. He borrows the money from a bank that charges 4% simple interest per year.

a. How much will he owe if he waits 1 year to pay back the loan? If he waits 2 years? 3 years? 4 years? 5 years?

b. Write a formula for how much he will owe after t years.

Example 2

Jack has $500 to invest. The bank offers an interest rate of 6% compounded annually. How much money will Jack have after 1 year? 2 years? 5 years? 10 years?

Example 3

If you have $200 to invest for 10 years, would you rather invest your money in a bank that pays 7% simple interest or in a bank that pays 5% interest compounded annually? Is there anything you could change in the problem that would make you change your answer?

EUREKA
MATH

Lesson Summary

SIMPLE INTEREST: Interest is calculated once per year on the original amount borrowed or invested. The interest does not become part of the amount borrowed or owed (the principal).

COMPOUND INTEREST: Interest is calculated once per period on the current amount borrowed or invested. Each period, the interest becomes a part of the principal.

Problem Set

1. $250 is invested at a bank that pays 7% simple interest. Calculate the amount of money in the account after 1 year, 3 years, 7 years, and 20 years.

2. $325 is borrowed from a bank that charges 4% interest compounded annually. How much is owed after 1 year, 3 years, 7 years, and 20 years?

3. Joseph has $10,000 to invest. He can go to Yankee Bank that pays 5% simple interest or Met Bank that pays 4% interest compounded annually. At how many years will Met Bank be the better choice?

This page intentionally left blank

Lesson 5: The Power of Exponential Growth

Classwork

Opening Exercise

Two equipment rental companies have different penalty policies for returning a piece of equipment late.

Company 1: On day 1, the penalty is $5. On day 2, the penalty is $10. On day 3, the penalty is $15. On day 4, the penalty is $20, and so on, increasing by $5 each day the equipment is late.

Company 2: On day 1, the penalty is $0.01. On day 2, the penalty is $0.02. On day 3, the penalty is $0.04. On day 4, the penalty is $0.08, and so on, doubling in amount each additional day late.

Jim rented a digger from Company 2 because he thought it had the better late return policy. The job he was doing with the digger took longer than he expected, but it did not concern him because the late penalty seemed so reasonable. When he returned the digger 15 days late, he was shocked by the penalty fee. What did he pay, and what would he have paid if he had used Company 1 instead?

Company 1		Company 2	
Day	**Penalty**	**Day**	**Penalty**
1		1	
2		2	
3		3	
4		4	
5		5	
6		6	
7		7	
8		8	
9		9	
10		10	
11		11	
12		12	
13		13	
14		14	
15		15	

a. Which company has a greater 15-day late charge?

b. Describe how the amount of the late charge changes from any given day to the next successive day in both Companies 1 and 2.

c. How much would the late charge have been after 20 days under Company 2?

Example 1

Folklore suggests that when the creator of the game of chess showed his invention to the country's ruler, the ruler was highly impressed. He was so impressed, he told the inventor to name a prize of his choice. The inventor, being rather clever, said he would take a grain of rice on the first square of the chessboard, two grains of rice on the second square of the chessboard, four on the third square, eight on the fourth square, and so on, doubling the number of grains of rice for each successive square. The ruler was surprised, even a little offended, at such a modest prize, but he ordered his treasurer to count out the rice.

a. Why is the ruler surprised? What makes him think the inventor requested a modest prize?

The treasurer took more than a week to count the rice in the ruler's store, only to notify the ruler that it would take more rice than was available in the entire kingdom. Shortly thereafter, as the story goes, the inventor became the new king.

b. Imagine the treasurer counting the needed rice for each of the 64 squares. We know that the first square is assigned a single grain of rice, and each successive square is double the number of grains of rice of the previous square. The following table lists the first five assignments of grains of rice to squares on the board. How can we represent the grains of rice as exponential expressions?

Square #	Grains of Rice	Exponential Expression
1	1	
2	2	
3	4	
4	8	
5	16	

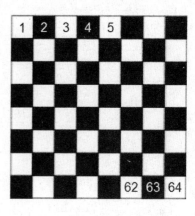

c. Write the exponential expression that describes how much rice is assigned to each of the last three squares of the board.

Square #	Exponential Expression
62	
63	
64	

Example 2

Let us understand the difference between $f(n) = 2n$ and $f(n) = 2^n$.

a. Complete the tables below, and then graph the points $\big(n, f(n)\big)$ on a coordinate plane for each of the formulas.

n	$f(n) = 2n$
-2	
-1	
0	
1	
2	
3	

n	$f(n) = 2^n$
-2	
-1	
0	
1	
2	
3	

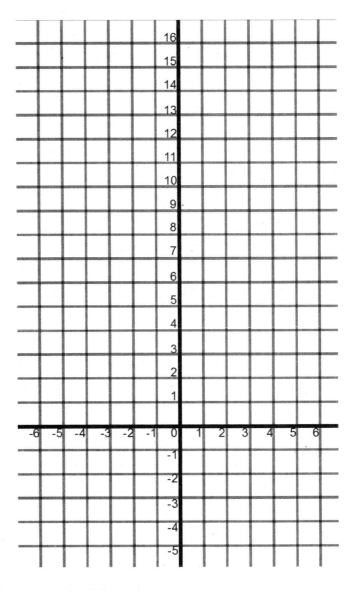

b. Describe the change in each sequence when n increases by 1 unit for each sequence.

EUREKA
MATH™

Exercise 1

A typical thickness of toilet paper is 0.001 inch. This seems pretty thin, right? Let's see what happens when we start folding toilet paper.

    a.   How thick is the stack of toilet paper after 1 fold? After 2 folds? After 5 folds?

    b.   Write an explicit formula for the sequence that models the thickness of the folded toilet paper after n folds.

    c.   After how many folds does the stack of folded toilet paper pass the 1-foot mark?

    d.   The moon is about 240,000 miles from Earth. Compare the thickness of the toilet paper folded 50 times to the distance from Earth.

Watch the following video *How folding paper can get you to the moon* (http://www.youtube.com/watch?v=AmFMJC45f1Q).

Exercise 2

A rare coin appreciates at a rate of 5.2% a year. If the initial value of the coin is $500, after how many years will its value cross the $3,000 mark? Show the formula that models the value of the coin after t years.

Problem Set

1. A bucket is put under a leaking ceiling. The amount of water in the bucket doubles every minute. After 8 minutes, the bucket is full. After how many minutes is the bucket half-full?

2. A three-bedroom house in Burbville sold for $190,000. If housing prices are expected to increase 1.8% annually in that town, write an explicit formula that models the price of the house in t years. Find the price of the house in 5 years.

3. A local college has increased its number of graduates by a factor of 1.045 over the previous year for every year since 1999. In 1999, 924 students graduated. What explicit formula models this situation? Approximately how many students will graduate in 2014?

4. The population growth rate of New York City has fluctuated tremendously in the last 200 years, the highest rate estimated at 126.8% in 1900. In 2001, the population of the city was 8,008,288, up 2.1% from 2000. If we assume that the annual population growth rate stayed at 2.1% from the year 2000 onward, in what year would we expect the population of New York City to have exceeded ten million people? Be sure to include the explicit formula you use to arrive at your answer.

5. In 2013, a research company found that smartphone shipments (units sold) were up 32.7% worldwide from 2012, with an expectation for the trend to continue. If 959 million units were sold in 2013, how many smartphones can be expected to sell in 2018 at the same growth rate? (Include the explicit formula for the sequence that models this growth.) Can this trend continue?

6. Two band mates have only 7 days to spread the word about their next performance. Jack thinks they can each pass out 100 fliers a day for 7 days, and they will have done a good job in getting the news out. Meg has a different strategy. She tells 10 of her friends about the performance on the first day and asks each of her 10 friends to each tell a friend on the second day and then everyone who has heard about the concert to tell a friend on the third day, and so on, for 7 days. Make an assumption that students are not telling someone who has not already been told.

   a. Over the first 7 days, Meg's strategy will reach fewer people than Jack's. Show that this is true.

   b. If they had been given more than 7 days, would there be a day on which Meg's strategy would begin to inform more people than Jack's strategy? If not, explain why not. If so, on which day would this occur?

   c. Knowing that she has only 7 days, how can Meg alter her strategy to reach more people than Jack does?

7. On June 1, a fast-growing species of algae is accidentally introduced into a lake in a city park. It starts to grow and cover the surface of the lake in such a way that the area it covers doubles every day. If it continues to grow unabated, the lake will be totally covered, and the fish in the lake will suffocate. At the rate it is growing, this will happen on June 30.

    a. When will the lake be covered halfway?

    b. On June 26, a pedestrian who walks by the lake every day warns that the lake will be completely covered soon. Her friend just laughs. Why might her friend be skeptical of the warning?

    c. On June 29, a cleanup crew arrives at the lake and removes almost all of the algae. When they are done, only 1% of the surface is covered with algae. How well does this solve the problem of the algae in the lake?

    d. Write an explicit formula for the sequence that models the percentage of the surface area of the lake that is covered in algae, a, given the time in days, t, that has passed since the algae was introduced into the lake.

8. Mrs. Davis is making a poster of math formulas for her students. She takes the 8.5 in. × 11 in. paper she printed the formulas on to the photocopy machine and enlarges the image so that the length and the width are both 150% of the original. She enlarges the image a total of 3 times before she is satisfied with the size of the poster. Write an explicit formula for the sequence that models the area of the poster, A, after n enlargements. What is the area of the final image compared to the area of the original, expressed as a percent increase and rounded to the nearest percent?

This page intentionally left blank

Lesson 6: Exponential Growth—U.S. Population and World Population

Mathematical Modeling Exercise 1

Callie and Joe are examining the population data in the graphs below for a history report. Their comments are as follows:

Callie: It looks like the U.S. population grew the same amount as the world population, but that can't be right, can it?

Joe: Well, I don't think they grew by the same *amount*, but they sure grew at about the same rate. Look at the slopes.

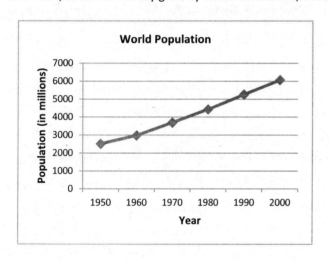

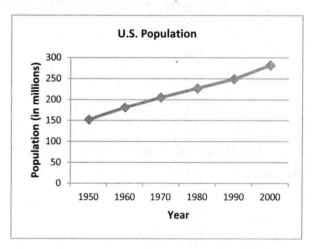

a. Is Callie's observation correct? Why or why not?

b. Is Joe's observation correct? Why or why not?

c. Use the World Population graph to estimate the percent increase in world population from 1950 to 2000.

d. Now, use the U.S. Population graph to estimate the percent increase in the U.S. population for the same time period.

e. How does the percent increase for the world population compare to that for the U.S. population over the same time period, 1950 to 2000?

f. Do the graphs above seem to indicate linear or exponential population growth? Explain your response.

g. Write an explicit formula for the sequence that models the world population growth from 1950 to 2000 based on the information in the graph. Assume that the population (in millions) in 1950 was 2,500 and in 2000 was 6,000. Use t to represent the number of years after 1950.

Mathematical Modeling Exercise 2

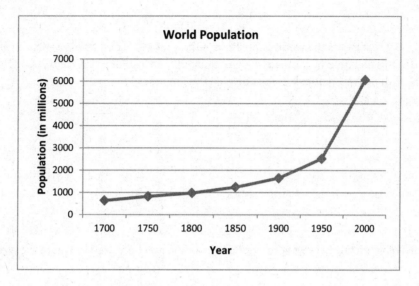

a.   How is this graph similar to the World Population graph in Mathematical Modeling Exercise 1? How is it different?

b.   Does the behavior of the graph from 1950 to 2000 match that shown on the graph in Mathematical Modeling Exercise 1?

c.   Why is the graph from Mathematical Modeling Exercise 1 somewhat misleading?

d. An exponential formula that can be used to model the world population growth from 1950 through 2000 is as follows:

$$f(t) = 2519(1.0177^t)$$

where 2,519 represents the world population in the year 1950, and t represents the number of years after 1950. Use this equation to calculate the world population in 1950, 1980, and 2000. How do your calculations compare with the world populations shown on the graph?

e. The following is a table showing the world population numbers used to create the graphs above.

Year	World Population (in millions)
1700	640
1750	824
1800	978
1850	1,244
1900	1,650
1950	2,519
1960	2,982
1970	3,692
1980	4,435
1990	5,263
2000	6,070

How do the numbers in the table compare with those you calculated in part (d) above?

f.   How is the formula in part (d) above different from the formula in Mathematical Modeling Exercise 1, part (g)? What causes the difference? Which formula more closely represents the population?

Exercises 1–2

1.   The table below represents the population of the United States (in millions) for the specified years.

Year	U.S. Population (in millions)
1800	5
1900	76
2000	282

a.   If we use the data from 1800 to 2000 to create an exponential equation representing the population, we generate the following formula for the sequence, where $f(t)$ represents the U.S. population and t represents the number of years after 1800.

$$f(t) = 5(1.0204)^t$$

Use this formula to determine the population of the United States in the year 2010.

b.   If we use the data from 1900 to 2000 to create an exponential equation that models the population, we generate the following formula for the sequence, where $f(t)$ represents the U.S. population and t represents the number of years after 1900.

$$f(t) = 76(1.013)^t$$

Use this formula to determine the population of the United States in the year 2010.

c. The actual U.S. population in the year 2010 was 309 million. Which of the above formulas better models the U.S. population for the entire span of 1800–2010? Why?

d. Complete the table below to show projected population figures for the years indicated. Use the formula from part (b) to determine the numbers.

Year	World Population (in millions)
2020	
2050	
2080	

e. Are the population figures you computed reasonable? What other factors need to be considered when projecting population?

2. The population of the country of Oz was 600,000 in the year 2010. The population is expected to grow by a factor of 5% annually. The annual food supply of Oz is currently sufficient for a population of 700,000 people and is increasing at a rate that will supply food for an additional 10,000 people per year.

a. Write a formula to model the population of Oz. Is your formula linear or exponential?

EUREKA
MATH™

b.   Write a formula to model the food supply. Is the formula linear or exponential?

c.   At what point does the population exceed the food supply? Justify your response.

d.   If Oz doubled its current food supply (to 1.4 million), would shortages still take place? Explain.

e.   If Oz doubles both its beginning food supply and doubles the rate at which the food supply increases, would food shortages still take place? Explain.

Problem Set

1. Student Friendly Bank pays a simple interest rate of 2.5% per year. Neighborhood Bank pays a compound interest rate of 2.1% per year, compounded monthly.

   a. Which bank will provide the largest balance if you plan to invest $10,000 for 10 years? For 20 years?

   b. Write an explicit formula for the sequence that models the balance in the Student Friendly Bank account t years after a deposit is left in the account.

   c. Write an explicit formula for the sequence that models the balance in the Neighborhood Bank account m months after a deposit is left in the account.

   d. Create a table of values indicating the balances in the two bank accounts from year 2 to year 20 in 2-year increments. Round each value to the nearest dollar.

Year	Student Friendly Bank (in dollars)	Neighborhood Bank (in dollars)
0		
2		
4		
6		
8		
10		
12		
14		
16		
18		
20		

   e. Which bank is a better short-term investment? Which bank is better for those leaving money in for a longer period of time? When are the investments about the same?

   f. What type of model is Student Friendly Bank? What is the rate or ratio of change?

   g. What type of model is Neighborhood Bank? What is the rate or ratio of change?

2. The table below represents the population of the state of New York for the years 1800–2000. Use this information to answer the questions.

Year	Population
1800	300,000
1900	7,300,000
2000	19,000,000

a. Using the year 1800 as the base year, an explicit formula for the sequence that models the population of New York is $P(t) = 300\,000(1.021)^t$, where t is the number of years after 1800.

Using this formula, calculate the projected population of New York in 2010.

b. Using the year 1900 as the base year, an explicit formula for the sequence that models the population of New York is $P(t) = 7\,300\,000(1.0096)^t$, where t is the number of years after 1900.

Using this formula, calculate the projected population of New York in 2010.

c. Using the Internet (or some other source), find the population of the state of New York according to the 2010 census. Which formula yielded a more accurate prediction of the 2010 population?

This page intentionally left blank

Lesson 7: Exponential Decay

Classwork

Example 1

a.   Malik bought a new car for $15,000. As he drove it off the lot, his best friend, Will, told him that the car's
     value just dropped by 15% and that it would continue to depreciate 15% of its current value each year. If the
     car's value is now $12,750 (according to Will), what will its value be after 5 years?

Complete the table below to determine the car's value after each of the next five years. Round each value to
the nearest cent.

Number of years, t, passed since driving the car off the lot	Car value after t years	15% depreciation of current car value	Car value minus the 15% depreciation
0	$12,750.00	$1,912.50	$10,837.50
1	10,837.50		
2			
3			
4			
5			

b.   Write an explicit formula for the sequence that models the value of Malik's car t years after driving it off the
     lot.

c.   Use the formula from part (b) to determine the value of Malik's car five years after its purchase. Round your
     answer to the nearest cent. Compare the value with the value in the table. Are they the same?

d.   Use the formula from part (b) to determine the value of Malik's car 7 years after its purchase. Round your
     answer to the nearest cent.

Exercises 1–6

1. Identify the initial value in each formula below, and state whether the formula models exponential growth or exponential decay. Justify your responses.

    a. $f(t) = 2 \left(\dfrac{2}{5}\right)^t$

    b. $f(t) = 2 \left(\dfrac{5}{3}\right)^t$

    c. $f(t) = \dfrac{2}{3}(3)^t$

    d. $f(t) = \dfrac{2}{3}\left(\dfrac{1}{3}\right)^t$

    e. $f(t) = \dfrac{3}{2}\left(\dfrac{2}{3}\right)^t$

2. If a person takes a given dosage d of a particular medication, then the formula $f(t) = d\,(0.8)^t$ represents the concentration of the medication in the bloodstream t hours later. If Charlotte takes 200 mg of the medication at $6:00$ a.m., how much remains in her bloodstream at $10:00$ a.m.? How long does it take for the concentration to drop below 1 mg?

EUREKA
MATH™

3. When you breathe normally, about 12% of the air in your lungs is replaced with each breath. Write an explicit formula for the sequence that models the amount of the original air left in your lungs, given that the initial volume of air is 500 ml. Use your model to determine how much of the original 500 ml remains after 50 breaths.

4. Ryan bought a new computer for $2,100. The value of the computer decreases by 50% each year. When will the value drop below $300?

5. Kelli's mom takes a 400 mg dose of aspirin. Each hour, the amount of aspirin in a person's system decreases by about 29%. How much aspirin is left in her system after 6 hours?

6. According to the International Basketball Association (FIBA), a basketball must be inflated to a pressure such that when it is dropped from a height of 1,800 mm, it rebounds to a height of 1,300 mm. Maddie decides to test the rebound-ability of her new basketball. She assumes that the ratio of each rebound height to the previous rebound height remains the same at $\frac{1300}{1800}$. Let $f(n)$ be the height of the basketball after n bounces. Complete the chart below to reflect the heights Maddie expects to measure.

n	$f(n)$
0	1,800
1	
2	
3	
4	

a.   Write the explicit formula for the sequence that models the height of Maddie's basketball after any number of bounces.

b.   Plot the points from the table. Connect the points with a smooth curve, and then use the curve to estimate the bounce number at which the rebound height drops below 200 mm.

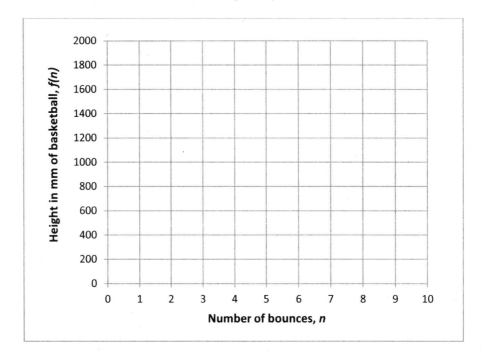

Lesson 7:   Exponential Decay

EUREKA
MATH™

Lesson Summary

The explicit formula $f(t) = ab^t$ models exponential decay, where a represents the initial value of the sequence, $b < 1$ represents the growth factor (or decay factor) per unit of time, and t represents units of time.

Problem Set

1. From 2000 to 2013, the value of the U.S. dollar has been shrinking. The value of the U.S. dollar over time $(v(t))$ can be modeled by the following formula:

    $$v(t) = 1.36(0.9758)^t, \text{ where } t \text{ is the number of years since 2000}$$

    a. How much was a dollar worth in the year 2005?

    b. Graph the points $(t, v(t))$ for integer values of $0 \le t \le 14$.

    c. Estimate the year in which the value of the dollar fell below $1.00.

2. A construction company purchased some equipment costing $300,000. The value of the equipment depreciates (decreases) at a rate of 14% per year.

    a. Write a formula that models the value of the equipment each year.

    b. What is the value of the equipment after 9 years?

    c. Graph the points $(t, v(t))$ for integer values of $0 \le t \le 15$.

    d. Estimate when the equipment will have a value of $50,000.

3. The number of newly reported cases of HIV (in thousands) in the United States from 2000 to 2010 can be modeled by the following formula:

    $$f(t) = 41(0.9842)^t, \text{ where } t \text{ is the number of years after 2000}$$

    a. Identify the growth factor.

    b. Calculate the estimated number of new HIV cases reported in 2004.

    c. Graph the points $(t, f(t))$ for integer values of $0 \le t \le 10$.

    d. During what year did the number of newly reported HIV cases drop below 36,000?

4. Doug drank a soda with 130 mg of caffeine. Each hour, the caffeine in the body diminishes by about 12%.

    a. Write a formula to model the amount of caffeine remaining in Doug's system each hour.

    b. How much caffeine remains in Doug's system after 2 hours?

    c. How long will it take for the level of caffeine in Doug's system to drop below 50 mg?

5.   64 teams participate in a softball tournament in which half the teams are eliminated after each round of play.

   a.   Write a formula to model the number of teams remaining after any given round of play.

   b.   How many teams remain in play after 3 rounds?

   c.   How many rounds of play will it take to determine which team wins the tournament?

6.   Sam bought a used car for $8,000. He boasted that he got a great deal since the value of the car two years ago (when it was new) was $15,000. His friend, Derek, was skeptical, stating that the value of a car typically depreciates about 25% per year, so Sam got a bad deal.

   a.   Use Derek's logic to write a formula for the value of Sam's car. Use t for the total age of the car in years.

   b.   Who is right, Sam or Derek?

EUREKA
MATH™

Lesson 8: Why Stay with Whole Numbers?

Opening Exercise

The sequence of perfect squares $\{1,4,9,16,25, \dots \}$ earned its name because the ancient Greeks realized these quantities could be arranged to form square shapes.

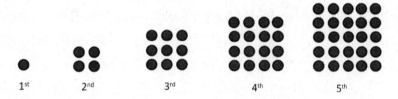

| 1st | 2nd | 3rd | 4th | 5th |

If $S(n)$ denotes the nth square number, what is a formula for $S(n)$?

Exercises

1. Prove whether or not 169 is a perfect square.

2. Prove whether or not 200 is a perfect square.

3. If $S(n) = 225$, then what is n?

4. Which term is the number 400 in the sequence of perfect squares? How do you know?

Instead of arranging dots into squares, suppose we extend our thinking to consider squares of side length x cm.

5. Create a formula for the area $A(x)$ cm^2 of a square of side length x cm: $A(x) = $ _____.

6. Use the formula to determine the area of squares with side lengths of 3 cm, 10.5 cm, and π cm.

7. What does $A(0)$ mean?

8. What does $A(-10)$ and $A(\sqrt{2})$ mean?

EUREKA
MATH™

The triangular numbers are the numbers that arise from arranging dots into triangular figures as shown:

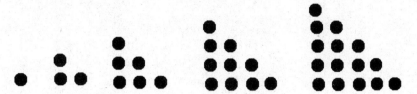

9. What is the 100th triangular number?

10. Find a formula for $T(n)$, the nth triangular number (starting with $n = 1$).

11. How can you be sure your formula works?

12. Create a graph of the sequence of triangular numbers $(n) = \frac{n(n+1)}{2}$, where n is a positive integer.

13. Create a graph of the triangle area formula $T(x) = \frac{x(x+1)}{2}$, where x is any positive real number.

14. How are your two graphs alike? How are they different?

Lesson 8:   Why Stay With Whole Numbers?

EUREKA
MATH™

Problem Set

1. The first four terms of two different sequences are shown below. Sequence A is given in the table, and sequence B is graphed as a set of ordered pairs.

n	$A(n)$
1	15
2	31
3	47
4	63

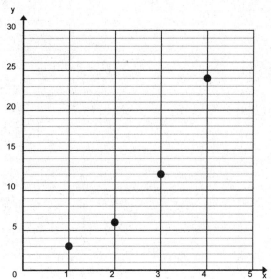

a. Create an explicit formula for each sequence.

b. Which sequence will be the first to exceed 500? How do you know?

2. A tile pattern is shown below.

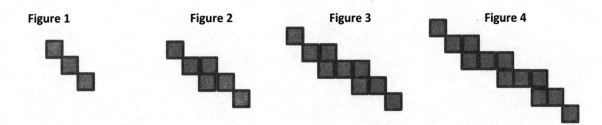

Figure 1 Figure 2 Figure 3 Figure 4

a. How is this pattern growing?

b. Create an explicit formula that could be used to determine the number of squares in the nth figure.

c. Evaluate your formula for $n = 0$, and $n = 2.5$. Draw Figure 0 and Figure 2.5, and explain how you decided to create your drawings.

3. The first four terms of a geometric sequence are graphed as a set of ordered pairs.

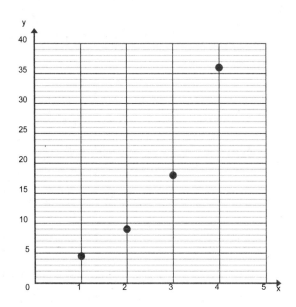

a. What is an explicit formula for this sequence?

b. Explain the meaning of the ordered pair $(3, 18)$.

c. As of July 2013, Justin Bieber had over 42,000,000 Twitter followers. Suppose the sequence represents the number of people that follow your new Twitter account each week since you started tweeting. If your followers keep growing in the same manner, when will you exceed 1,000,000 followers?

EUREKA
MATH™

Lesson 9: Representing, Naming, and Evaluating Functions

Classwork

Opening Exercise

Match each picture to the correct word by drawing an arrow from the word to the picture.

| Elephant |
| Camel |
| Polar Bear |
| Zebra |

FUNCTION: A *function* is a correspondence between two sets, X and Y, in which each element of X is matched to one and only one element of Y. The set X is called the *domain of the function*.

The notation $f: X \to Y$ is used to name the function and describes both X and Y. If x is an element in the domain X of a function $f: X \to Y$, then x is matched to an element of Y called $f(x)$. We say $f(x)$ is the value in Y that denotes the *output* or *image* of f corresponding to the *input* x.

The *range (or image)* of a function $f: X \to Y$ is the subset of Y, denoted $f(X)$, defined by the following property: y is an element of $f(X)$ if and only if there is an x in X such that $f(x) = y$.

Example 1

Define the Opening Exercise using function notation. State the domain and the range.

Example 2

Is the assignment of students to English teachers an example of a function? If yes, define it using function notation, and state the domain and the range.

Example 3

Let $X = \{1, 2, 3, 4\}$ and $Y = \{5, 6, 7, 8, 9\}$. f and g are defined below.

$f: X \to Y$ $g: X \to Y$

$f = \{(1,7), (2,5), (3,6), (4,7)\}$ $g = \{(1,5), (2,6), (1,8), (2,9), (3,7)\}$

Is f a function? If yes, what is the domain, and what is the range? If no, explain why f is not a function.

Is g a function? If yes, what is the domain and range? If no, explain why g is not a function.

What is $f(2)$?

If $f(x) = 7$, then what might x be?

Lesson 9:   Representing, Naming, and Evaluating Functions

Exercises

1. Define f to assign each student at your school a unique ID number.

    f: {students in your school} → {whole numbers}

    Assign each student a unique ID number.

    a. Is this an example of a function? Use the definition to explain why or why not.

    b. Suppose $f(\text{Hilda}) = 350123$. What does that mean?

    c. Write your name and student ID number using function notation.

2. Let g assign each student at your school to a grade level.

    a. Is this an example of a function? Explain your reasoning.

    b. Express this relationship using function notation, and state the domain and the range.

    g: {students in the school} → {grade level}

    Assign each student to a grade level.

3. Let h be the function that assigns each student ID number to a grade level.

$h: \{$student ID number$\} \to \{$grade level$\}$

Assign each student ID number to the student's current grade level.

a. Describe the domain and range of this function.

b. Record several ordered pairs $(x, f(x))$ that represent yourself and students in your group or class.

c. Jonny says, "This is not a function because every ninth grader is assigned the same range value of 9. The range only has 4 numbers $\{9, 10, 11, 12\}$, but the domain has a number for every student in our school." Explain to Jonny why he is incorrect.

EUREKA
MATH™

Problem Set

1. Which of the following are examples of a function? Justify your answers.

   a. The assignment of the members of a football team to jersey numbers.

   b. The assignment of U.S. citizens to Social Security numbers.

   c. The assignment of students to locker numbers.

   d. The assignment of the residents of a house to the street addresses.

   e. The assignment of zip codes to residences.

   f. The assignment of residences to zip codes.

   g. The assignment of teachers to students enrolled in each of their classes.

   h. The assignment of all real numbers to the next integer equal to or greater than the number.

   i. The assignment of each rational number to the product of its numerator and denominator.

2. Sequences are functions. The domain is the set of all term numbers (which is usually the positive integers), and the range is the set of terms of the sequence. For example, the sequence 1, 4, 9, 16, 25, 36, … of perfect squares is the function:

   Let f: {positive integers} $\rightarrow$ {perfect squares}

   Assign each term number to the square of that number.

   a. What is $f(3)$? What does it mean?

   b. What is the solution to the equation $f(x) = 49$? What is the meaning of this solution?

   c. According to this definition, is -3 in the domain of f? Explain why or why not.

   d. According to this definition, is 50 in the range of f? Explain why or why not.

3. Write each sequence as a function.

   a. $\{1, 3, 6, 10, 15, 21, 28\}$

   b. $\{1, 3, 5, 7, 9, …\}$

   c. $a_{n+1} = 3a_n$, $a_1 = 1$, where n is a positive integer greater than or equal to 1.

This page intentionally left blank

Lesson 10: Representing, Naming, and Evaluating Functions

Classwork

Opening Exercise

Study the 4 representations of a function below. How are these representations alike? How are they different?

TABLE:

Input	0	1	2	3	4	5
Output	1	2	4	8	16	32

FUNCTION:

Let $f: \{0, 1, 2, 3, 4, 5\} \to \{1, 2, 4, 8, 16, 32\}$ such that $x \mapsto 2^x$.

SEQUENCE:

Let $a_{n+1} = 2a_n, a_0 = 1$ for $0 \le n \le 4$ where n is an integer.

DIAGRAM:

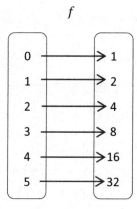

Exercise 1

Let $X = \{0, 1, 2, 3, 4, 5\}$. Complete the following table using the definition of f.

$f: X \rightarrow Y$

Assign each x in X to the expression 2^x.

x	0	1	2	3	4	5
$f(x)$						

What are $f(0), f(1), f(2), f(3), f(4)$, and $f(5)$?

What is the range of f?

Exercise 2

The squaring function is defined as follows:

Let $f: X \rightarrow Y$ be the function such that $x \mapsto x^2$, where X is the set of all real numbers.

What are $f(0), f(3), f(-2), f(\sqrt{3}), f(-2.5), f\left(\frac{2}{3}\right), f(a)$, and $f(3 + a)$?

What is the range of f?

Lesson 10:   Representing, Naming, and Evaluating Functions

EUREKA
MATH™

What subset of the real numbers could be used as the domain of the squaring function to create a range with the same output values as the sequence of square numbers $\{1, 4, 9, 16, 25, \dots\}$ from Lesson 9?

Exercise 3

Recall that an equation can either be true or false. Using the function defined by $f : \{0, 1, 2, 3, 4, 5\} \to \{1, 2, 4, 8, 16, 32\}$ such that $x \mapsto 2^x$, determine whether the equation $f(x) = 2^x$ is true or false for each x in the domain of f.

x	Is the equation $f(x) = 2^x$ true or false?	Justification
0	True	Substitute 0 into the equation. $f(0) = 2^0$ $1 = 2^0$ The 1 on the left side comes from the definition of f, and the value of 2^0 is also 1, so the equation is true.
1		
2		
3		
4		
5		

If the domain of f were extended to all real numbers, would the equation still be true for each x in the domain of f? Explain your thinking.

Exercise 4

Write three different polynomial functions such that $f(3) = 2$.

Exercise 5

The domain and range of this function are not specified. Evaluate the function for several values of x. What subset of the real numbers would represent the domain of this function? What subset of the real numbers would represent its range?

$$\text{Let } f(x) = \sqrt{x - 2}$$

EUREKA
MATH™

Lesson Summary

ALGEBRAIC FUNCTION: Given an algebraic expression in one variable, an *algebraic function* is a function $f: D \to Y$ such that for each real number x in the domain D, $f(x)$ is the value found by substituting the number x into all instances of the variable symbol in the algebraic expression and evaluating.

The following notation will be used to define functions going forward. If a domain is not specified, it is assumed to be the set of all real numbers.

For the squaring function, we say   Let $f(x) = x^2$.

For the exponential function with base 2, we say   Let $f(x) = 2^x$.

When the domain is limited by the expression or the situation to be a subset of the real numbers, it must be specified when the function is defined.

For the square root function, we say   Let $f(x) = \sqrt{x}$ for $x \geq 0$.

To define the first 5 triangular numbers, we say   Let $f(x) = \dfrac{x(x+1)}{2}$ for $1 \leq x \leq 5$ where x is an integer.

Depending on the context, one either views the statement "$f(x) = \sqrt{x}$" as part of defining the function f or as an equation that is true for all x in the domain of f or as a formula.

Problem Set

1. Let $f(x) = 6x - 3$, and let $g(x) = 0.5(4)^x$. Find the value of each function for the given input.

   a. $f(0)$

   b. $f(-10)$

   c. $f(2)$

   d. $f(0.01)$

   e. $f(11.25)$

   f. $f(-\sqrt{2})$

   g. $f\left(\dfrac{5}{3}\right)$

   h. $f(1) + f(2)$

   i. $f(6) - f(2)$

   j. $g(0)$

   k. $g(-1)$

   l. $g(2)$

   m. $g(-3)$

   n. $g(4)$

   o. $g(\sqrt{2})$

   p. $g\left(\dfrac{1}{2}\right)$

   q. $g(2) + g(1)$

   r. $g(6) - g(2)$

2. Since a variable is a placeholder, we can substitute letters that stand for numbers in for x. Let $f(x) = 6x - 3$, and let $g(x) = 0.5(4)^x$, and suppose a, b, c, and h are real numbers. Find the value of each function for the given input.

    a. $f(a)$

    b. $f(2a)$

    c. $f(b + c)$

    d. $f(2 + h)$

    e. $f(a + h)$

    f. $f(a + 1) - f(a)$

    g. $f(a + h) - f(a)$

    h. $g(b)$

    i. $g(b + 3)$

    j. $g(3b)$

    k. $g(b - 3)$

    l. $g(b + c)$

    m. $g(b + 1) - g(b)$

3. What is the range of each function given below?

    a. Let $f(x) = 9x - 1$.

    b. Let $g(x) = 3^{2x}$.

    c. Let $f(x) = x^2 - 4$.

    d. Let $h(x) = \sqrt{x} + 2$.

    e. Let $a(x) = x + 2$ such that x is a positive integer.

    f. Let $g(x) = 5^x$ for $0 \le x \le 4$.

4. Provide a suitable domain and range to complete the definition of each function.

    a. Let $f(x) = 2x + 3$.

    b. Let $f(x) = 2^x$.

    c. Let $C(x) = 9x + 130$, where $C(x)$ is the number of calories in a sandwich containing x grams of fat.

    d. Let $B(x) = 100(2)^x$, where $B(x)$ is the number of bacteria at time x hours over the course of one day.

5. Let $f : X \rightarrow Y$, where X and Y are the set of all real numbers, and x and h are real numbers.

    a. Find a function f such that the equation $f(x + h) = f(x) + f(h)$ is not true for all values of x and h. Justify your reasoning.

    b. Find a function f such that equation $f(x + h) = f(x) + f(h)$ is true for all values of x and h. Justify your reasoning.

    c. Let $f(x) = 2^x$. Find a value for x and a value for h that makes $f(x + h) = f(x) + f(h)$ a true number sentence.

EUREKA
MATH™

6. Given the function f whose domain is the set of real numbers, let $f(x) = 1$ if x is a rational number, and let $f(x) = 0$ if x is an irrational number.

a. Explain why f is a function.

b. What is the range of f?

c. Evaluate f for each domain value shown below.

x	$\dfrac{2}{3}$	0	-5	$\sqrt{2}$	π
$f(x)$					

d. List three possible solutions to the equation $f(x) = 0$.

This page intentionally left blank

Lesson 11: The Graph of a Function

Classwork

In Module 1, you graphed equations such as $y = 10 - 4x$ by plotting the points in the Cartesian plane by picking x-values and then using the equation to find the y-value for each x-value. The number of ordered pairs you plotted to get the general shape of the graph depended on the type of equation (linear, quadratic, etc.). The graph of the equation was then a representation of the solution set, which could be described using set notation.

In this lesson, we extend set notation slightly to describe the graph of a function. In doing so, we explain a way to think about set notation for the graph of a function that mimics the instructions a tablet or laptop might perform to draw a graph on its screen.

Exploratory Challenge 1

Computer programs are essentially instructions to computers on what to do when the user (you!) makes a request. For example, when you type a letter on your smart phone, the smart phone follows a specified set of instructions to draw that letter on the screen and record it in memory (as part of an email, for example). One of the simplest types of instructions a computer can perform is a *for-next loop*. Below is code for a program that prints the first 5 powers of 2:

```
Declare x integer
For all x from 1 to 5
        Print 2^x
Next x
```

The output of this program code is

2
4
8
16
32

Here is a description of the instructions: First, x is quantified as an integer, which means the variable can only take on integer values and cannot take on values like $\frac{1}{3}$ or $\sqrt{2}$. The *For* statement begins the loop, starting with $x = 1$. The instructions between *For* and *Next* are performed for the value $x = 1$, which in this case is just to *Print* 2. (Print means "print to the computer screen.") Then the computer performs the instructions again for the next x ($x = 2$), that is, *Print* 4, and so on until the computer performs the instructions for $x = 5$, that is, *Print* 32.

Exercise 1

Perform the instructions in the following programming code as if you were a computer and your paper was the computer screen.

```
Declare x integer
For all x from 2 to 8
        Print 2x + 3
Next x
```

Exploratory Challenge 2

We can use almost the same code to build a set: First, we start with a set with zero elements in it (called the *empty set*), and then we increase the size of the set by appending one new element to it in each for-next step.

```
Declare x integer
Initialize G as {}
For all x from 2 to 8
        Append 2x + 3 to G
        Print G
Next x
```

Note that G is printed to the screen after each new number is appended. Thus, the output shows how the set builds:

$\{7\}$
$\{7, 9\}$
$\{7, 9, 11\}$
$\{7, 9, 11, 13\}$
$\{7, 9, 11, 13, 15\}$
$\{7, 9, 11, 13, 15, 17\}$
$\{7, 9, 11, 13, 15, 17, 19\}.$

EUREKA
MATH

Exercise 2

We can also build a set by appending ordered pairs. Perform the instructions in the following programming code as if you were a computer and your paper were the computer screen (the first few are done for you).

```
Declare x integer
Initialize G as {}
For all x from 2 to 8
        Append (x, 2x + 3) to G
Next x
Print G
```

Output:
$\{(2,7), (3,9),$ _____ $\}$

Exploratory Challenge 3

Instead of Printing the set G to the screen, we can use another command, *Plot*, to plot the points on a Cartesian plane.

```
Declare x integer
Initialize G as {}
For all x from 2 to 8
        Append (x, 2x + 3) to G
Next x
Plot G
```

Output:

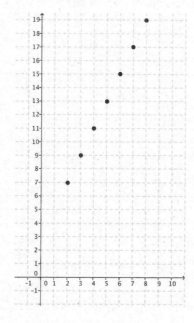

In mathematics, the programming code above can be compactly written using set notation, as follows:

$$\{(x, 2x + 3) \mid x \text{ integer} \text{ and } 2 \le x \le 8\}.$$

This set notation is an abbreviation for "The set of all points $(x, 2x + 3)$ such that x is an integer and $2 \le x \le 8$." Notice how the set of ordered pairs generated by the for-next code above,

$$\{(2,7), (3,9), (4,11), (5,13), (6,15), (7,17), (8,19)\},$$

also satisfies the requirements described by $\{(x, 2x + 3) \mid x \text{ integer}, 2 \le x \le 8\}$. It is for this reason that the set notation of the form

$$\{\text{type of element} \mid \text{condition on each element}\}$$

is sometimes called *set-builder notation*—because it can be thought of as building the set just like the for-next code.

Discussion

We can now upgrade our notion of a for-next loop by doing a thought experiment: Imagine a for-next loop that steps through *all* real numbers in an interval (not just the integers). No computer can actually do this—computers can only do a finite number of calculations. But our human brains are far superior to that of any computer, and we can easily imagine what that might look like. Here is some sample code:

```
Declare x real
Let f(x) = 2x + 3
Initialize G as {}
For all x such that 2 ≤ x ≤ 8
        Append (x, f(x)) to G
Next x
Plot G
```

The output of this thought code is the graph of f for all real numbers x in the interval $2 \le x \le 8$:

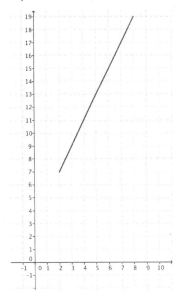

Lesson 11:   The Graph of a Function

Exercise 3

a.   Plot the function f on the Cartesian plane using the following for-next thought code.

> **Declare x real**
> **Let $f(x) = x^2 + 1$**
> **Initialize G as {}**
> **For all x such that $-2 \leq x \leq 3$**
>   **Append $(x, f(x))$ to G**
> **Next x**
> **Plot G**

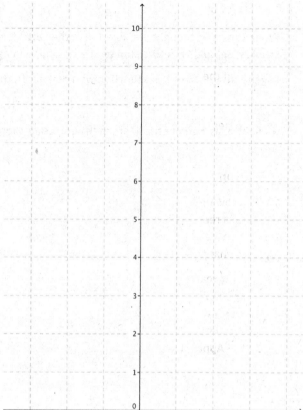

b.   For each step of the for-next loop, what is the input value?

c.   For each step of the for-next loop, what is the output value?

d.   What is the domain of the function f?

e.   What is the range of the function f?

Closing

The set G built from the for-next thought code in Exercise 4 can also be compactly written in mathematics using set notation:

$$\{(x, x^2 + 1) \mid x \text{ real}, -2 \le x \le 3\}.$$

When this set is thought of as plotted in the Cartesian plane, it is the same graph. When you see this set notation in the Problem Set and/or future studies, it is helpful to imagine this set-builder notation as describing a for-next loop.

In general, if $f: D \to Y$ is a function with domain D, then its *graph* is the set of all ordered pairs,

$$\{(x, f(x)) \mid x \in D\},$$

thought of as a geometric figure in the Cartesian coordinate plane. (The symbol $\in$ simply means "in." The statement $x \in D$ is read, "x in D.")

EUREKA MATH

> **Lesson Summary**
>
> **GRAPH OF f:** Given a function f whose domain D and range are subsets of the real numbers, the graph of f is the set of ordered pairs in the Cartesian plane given by
>
> $$\{(x, f(x)) \mid x \in D\}.$$

Problem Set

1. Perform the instructions for each of the following programming codes as if you were a computer and your paper was the computer screen.

   a.

   ```
   Declare x integer
   For all x from 0 to 4
           Print 2x
   Next x
   ```

   b.

   ```
   Declare x integer
   For all x from 0 to 10
           Print 2x + 1
   Next x
   ```

   c.

   ```
   Declare x integer
   For all x from 2 to 8
           Print x²
   Next x
   ```

   d.

   ```
   Declare x integer
   For all x from 0 to 4
           Print 10 · 3ˣ
   Next x
   ```

2. Perform the instructions for each of the following programming codes as if you were a computer and your paper were the computer screen.

   a.

   ```
   Declare x integer
   Let f(x) = (x + 1)(x − 1) − x²
   Initialize G as {}
   For all x from −3 to 3
           Append (x, f(x)) to G
   Next x
   Plot G
   ```

   b.

   ```
   Declare x integer
   Let f(x) = 3⁻ˣ
   Initialize G as {}
   For all x from −3 to 3
           Append (x, f(x)) to G
   Next x
   Plot G
   ```

   c.

   ```
   Declare x real
   Let f(x) = x³
   Initialize G as {}
   For all x such that −2 ≤ x ≤ 2
           Append (x, f(x)) to G
   Next x
   Plot G
   ```

EUREKA
MATH™

3. Answer the following questions about the thought code:

> **Declare x real**
> **Let $f(x) = (x - 2)(x - 4)$**
> **Initialize G as {}**
> **For all x such that $0 \leq x \leq 5$**
> **Append $(x, f(x))$ to G**
> **Next x**
> **Plot G**

a. What is the domain of the function f?

b. Plot the graph of f according to the instructions in the thought code.

c. Look at your graph of f. What is the range of f?

d. Write three or four sentences describing in words how the thought code works.

4. Sketch the graph of the functions defined by the following formulas, and write the graph of f as a set using set-builder notation. (Hint: Assume the domain is all real numbers unless specified in the problem.)

a. $f(x) = x + 2$

b. $f(x) = 3x + 2$

c. $f(x) = 3x - 2$

d. $f(x) = -3x - 2$

e. $f(x) = -3x + 2$

f. $f(x) = -\frac{1}{3}x + 2, -3 \leq x \leq 3$

g. $f(x) = (x + 1)^2 - x^2, -2 \leq x \leq 5$

h. $f(x) = (x + 1)^2 - (x - 1)^2, -2 \leq x \leq 4$

5. The figure shows the graph of $f(x) = -5x + c$.

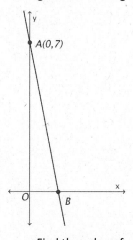

a. Find the value of c.

b. If the graph of f intersects the x-axis at B, find the coordinates of B.

6. The figure shows the graph of $f(x) = \frac{1}{2}x + c$.

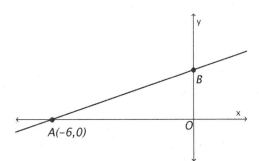

a. Find the value of c.

b. If the graph of f intersects the y-axis at B, find the coordinates of B.

c. Find the area of triangle AOB.

EUREKA
MATH

Lesson 12: The Graph of the Equation $y = f(x)$

Classwork

In Module 1, you graphed equations such as $4x + y = 10$ by plotting the points on the Cartesian coordinate plane that corresponded to all of the ordered pairs of numbers (x, y) that were in the solution set. We called the geometric figure that resulted from plotting those points in the plane the *graph of the equation in two variables*.

In this lesson, we extend this notion of the graph of an equation to the graph of $y = f(x)$ for a function f. In doing so, we use computer thought code to describe the process of generating the ordered pairs in the graph of $y = f(x)$.

Example 1

In the previous lesson, we studied a simple type of instruction that computers perform called a for-next loop. Another simple type of instruction is an *if-then statement*. Below is example code of a program that tests for and prints "True" when $x + 2 = 4$; otherwise it prints "False."

```
Declare x integer
For all x from 1 to 4
    If x + 2 = 4 then
        Print True
    else
        Print False
    End if
Next x
```

The output of this program code is

False
True
False
False

Notice that the if-then statement in the code above is really just testing whether each number in the loop is in the solution set.

Example 2

Perform the instructions in the following programming code as if you were a computer and your paper were the computer screen.

```
Declare x integer
Initialize G as {}
For all x from 0 to 4
    If x² − 4x + 5 = 2 then
        Append x to G
    else
        Do NOT append x to G
    End if
Next x
Print G
```

Output: $\{1, 3\}$

Discussion

Compare the for-next/if-then code above to the following set-builder notation we used to describe solution sets in Module 1:

$$\{x \text{ integer} \mid 0 \le x \le 4 \text{ and } x^2 - 4x + 5 = 2\}.$$

Check to see that the set-builder notation also generates the set $\{1, 3\}$. *Whenever you see set-builder notation to describe a set, a powerful way to interpret that notation is to think of the set as being generated by a program like the for-next or if-then code above.*

Exploratory Challenge 1

Next we write code that generates a graph of a *two-variable equation* $y = x(x - 2)(x + 2)$ for x in $\{-2, -1, 0, 1, 2\}$ and y in $\{-3, 0, 3\}$. The solution set of this equation is generated by testing each ordered pair (x, y) in the set,

$$\{(-2, -3), (-2,0), (-2,3), (-1, -3), (-1,0), (-1,3), \ldots, (2, -3), (2,0), (2,3)\},$$

to see if it is a solution to the equation $y = x(x - 2)(x + 2)$. Then the graph is just the plot of solutions in the Cartesian plane. We can instruct a computer to find these points and plot them using the following program.

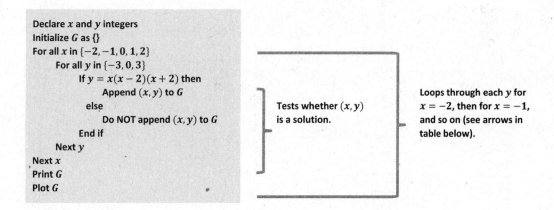

a. Use the table below to record the decisions a computer would make when following the program instructions above. Fill in each cell with "Yes" or "No" depending on whether the ordered pair (x, y) would be appended or not. (The step where $x = -2$ has been done for you.)

	$x = -2$	$x = -1$	$x = 0$	$x = 1$	$x = 2$
$y = 3$	No				
$y = 0$	Yes				
$y = -3$	No				

b. What would be the output to the Print G command? (The first ordered pair is listed for you.)

Output:
$\{ \ (-2,0) \quad , _____, _____, _____, _____ \}$

c. Plot the solution set G in the Cartesian plane. (The first ordered pair in G has been plotted for you.)

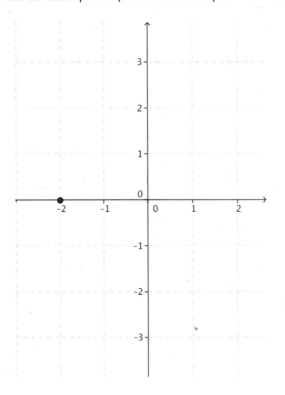

Exploratory Challenge 2

The program code in Exercise 3 is a way to imagine how set-builder notation generates solution sets and figures in the plane. Given a function $f(x) = x(x - 2)(x - 3)$ with domain and range all real numbers, a slight modification of the program code above can be used to generate the graph of the equation $y = f(x)$:

$$\{(x, y) \mid x \text{ real } \text{ and } y = f(x)\}.$$

Even though the code below cannot be run on a computer, students can run the following thought code in their minds.

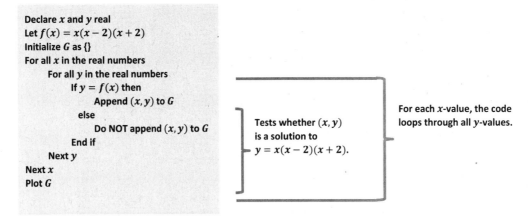

```
Declare x and y real
Let f(x) = x(x - 2)(x + 2)
Initialize G as {}
For all x in the real numbers
        For all y in the real numbers
            If y = f(x) then
                    Append (x, y) to G
            else
                    Do NOT append (x, y) to G
            End if
        Next y
Next x
Plot G
```

Tests whether (x, y) is a solution to $y = x(x - 2)(x + 2)$.

For each x-value, the code loops through all y-values.

a.   Plot G on the Cartesian plane (the figure drawn is called the graph of $y = f(x)$).

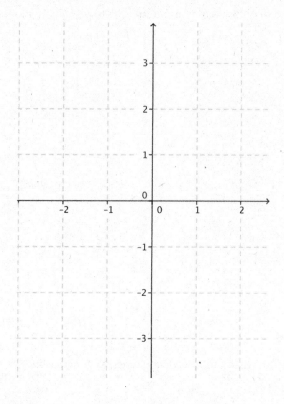

b.   Describe how the thought code is similar to the set-builder notation $\{(x, y) \mid x \text{ real and } y = f(x)\}$.

c.   A *relative maximum* for the function f occurs at the x-coordinate of $\left(-\frac{2}{3}\sqrt{3}, \frac{16}{9}\sqrt{3}\right)$. Substitute this point into the equation $y = x(x^2 - 4)$ to check that it is a solution to $y = f(x)$, and then plot the point on your graph.

d.   A *relative minimum* for the function f occurs at the x-coordinate of $\left(\frac{2}{3}\sqrt{3}, -\frac{16}{9}\sqrt{3}\right)$. A similar calculation as you did above shows that this point is also a solution to $y = f(x)$. Plot this point on your graph.

e.   Look at your graph. On what interval(s) is the function f decreasing?

f.   Look at your graph. On what interval(s) is the function f increasing?

EUREKA
MATH

Lesson Summary

- **GRAPH OF $y = f(x)$:** Given a function f whose domain D, and the range are subsets of the real numbers, the *graph of $y = f(x)$* is the set of ordered pairs (x, y) in the Cartesian plane given by

$$\{(x, y) \mid x \in D \text{ and } y = f(x)\}.$$

When we write $\{(x, y) \mid y = f(x)\}$ for the graph of $y = f(x)$, it is understood that the domain is the largest set of real numbers for which the function f is defined.

- The graph of f is the same as the graph of the equation $y = f(x)$.

- **INCREASING/DECREASING:** Given a function f whose domain and range are subsets of the real numbers, and I is an interval contained within the domain, the function is called *increasing on the interval I* if

$$f(x_1) < f(x_2) \text{ whenever } x_1 < x_2 \text{ in } I.$$

It is called *decreasing on the interval I* if

$$f(x_1) > f(x_2) \text{ whenever } x_1 < x_2 \text{ in } I.$$

Problem Set

1. Perform the instructions in the following programming code as if you were a computer and your paper were the computer screen.

```
Declare x integer
For all x from 1 to 6
    If x² − 2 = 7 then
        Print True
    else
        Print False
    End if
Next x
```

2. Answer the following questions about the computer programming code.

```
Declare x integer
Initialize G as {}
For all x from −3 to 3
    If 2^x + 2^−x = 17/4 then
        Append x to G
    else
        Do NOT append x to G
    End if
Next x
Print G
```

a. Perform the instructions in the programming code as if you were a computer and your paper were the computer screen.

b. Write a description of the set G using set-builder notation.

3. Answer the following questions about the computer programming code.

```
Declare x and y integers
Initialize G as {}
For all x in {0, 1, 2, 3}
    For all y in {0, 1, 2, 3}
        If y = √(4 + 20x − 19x² + 4x³) then
            Append (x, y) to G
        else
            Do NOT append (x, y) to G
        End if
    Next y
Next x
Plot G
```

a. Use the table below to record the decisions a computer would make when following the program instructions above. Fill in each cell with "Yes" or "No" depending on whether the ordered pair (x, y) would be appended or not.

	$x = 0$	$x = 1$	$x = 2$	$x = 3$
$y = 3$				
$y = 2$				
$y = 1$				
$y = 0$				

Lesson 12: The Graph of the Equation $y = f(x)$

EUREKA MATH™

b. Plot the set G in the Cartesian plane.

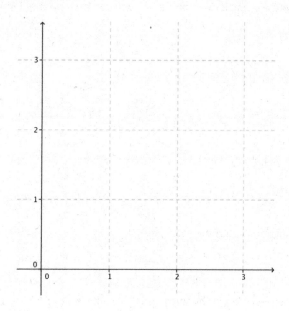

4. Answer the following questions about the thought code.

```
Declare x and y real
Let f(x) = −2x + 8
Initialize G as {}
For all x in the real numbers
    For all y in the real numbers
        If y = f(x) then
            Append (x, y) to G
        else
            Do NOT append (x, y) to G
        End if
    Next y
Next x
Plot G
```

a. What is the domain of the function $f(x) = -2x + 8$?

b. What is the range of the function $f(x) = -2x + 8$?

c. Write the set G generated by the thought code in set-builder notation.

d. Plot the set G to obtain the graph of the function $f(x) = -2x + 8$.

e. The function $f(x) = -2x + 8$ is clearly a decreasing function on the domain of the real numbers. Show that the function satisfies the definition of decreasing for the points 8 and 10 on the number line; that is, show that since $8 < 10$, then $f(8) > f(10)$.

5. Sketch the graph of the functions defined by the following formulas, and write the graph of $y = f(x)$ as a set using set-builder notation. (Hint: For each function below, you can assume the domain is all real numbers.)

   a. $f(x) = -\frac{1}{2}x + 6$

   b. $f(x) = x^2 + 3$

   c. $f(x) = x^2 - 5x + 6$

   d. $f(x) = x^3 - x$

   e. $f(x) = -x^2 + x - 1$

   f. $f(x) = (x - 3)^2 + 2$

   g. $f(x) = x^3 - 2x^2 + 3$

6. Answer the following questions about the set:

$$\{(x, y) \mid 0 \le x \le 2 \text{ and } y = 9 - 4x^2\}.$$

   a. The equation can be rewritten in the form $y = f(x)$ where $f(x) = 9 - 4x^2$. What are the domain and range of the function f specified by the set?

   i. Domain:

   ii. Range:

   b. Write thought code such as that in Problem 4 that will generate and then plot the set.

7. Answer the following about the graph of a function below.

   a. Which points (A, B, C, or D) are relative maxima?

   b. Which points (A, B, C, or D) are relative minima?

   c. Name any interval where the function is increasing.

   d. Name any interval where the function is decreasing.

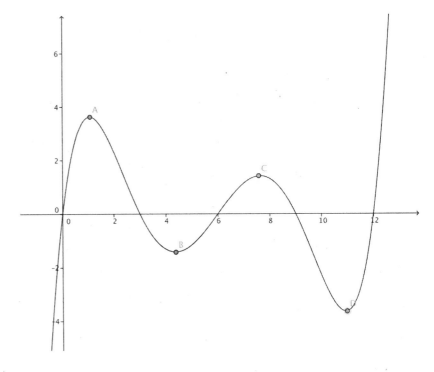

EUREKA
MATH™

Lesson 13: Interpreting the Graph of a Function

Classwork

This graphic was shared by NASA prior to the Mars Curiosity Rover landing on August 6, 2012. It depicts the landing sequence for the Curiosity Rover's descent to the surface of the planet.

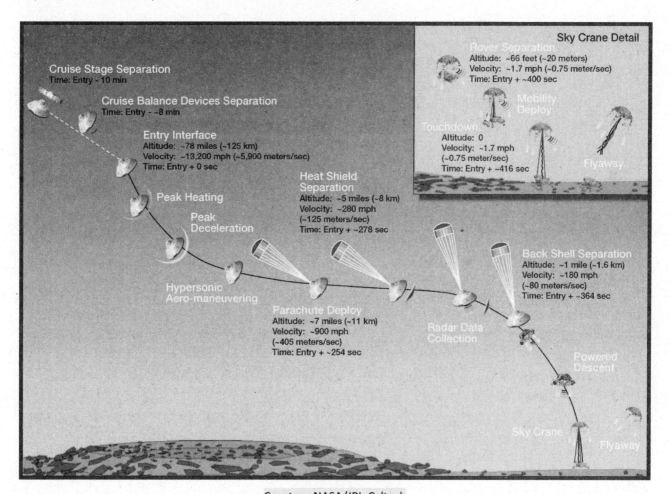

Courtesy NASA/JPL-Caltech

Does this graphic really represent the landing path of the Curiosity Rover? Create a model that can be used to predict the altitude and velocity of the Curiosity Rover 5, 4, 3, 2, and 1 minute before landing.

Mathematical Modeling Exercise

Create a model to help you answer the problem and estimate the altitude and velocity at various times during the landing sequence.

Lesson 13:   Interpreting the Graph of a Function

Exercises

1. Does this graphic really represent the landing path of the Curiosity Rover?

2. Estimate the altitude and velocity of the Curiosity Rover 5, 4, 3, 2, and 1 minute before landing. Explain how you arrived at your estimate.

3. Based on watching the video/animation, do you think you need to revise any of your work? Explain why or why not, and then make any needed changes.

4. Why is the graph of the altitude function decreasing and the graph of the velocity function increasing on its domain?

5. Why is the graph of the velocity function negative? Why does this graph not have a t-intercept?

6. What is the meaning of the t-intercept of the altitude graph? The y-intercept?

EUREKA
MATH™

A Mars rover collected the following temperature data over 1.6 Martian days. A Martian day is called a sol. Use the graph to answer the following questions.

GROUND AND AIR TEMPERATURE SENSOR

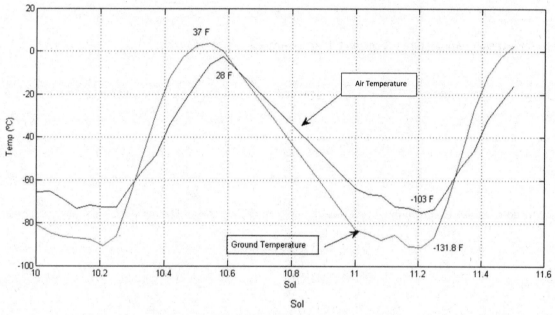

Courtesy NASA/JPL-Caltech/CAB(CSIC-INTA)

7. Approximately when does each graph change from increasing to decreasing? From decreasing to increasing?

8. When is the air temperature increasing?

9. When is the ground temperature decreasing?

10. What is the air temperature change on this time interval?

11. Why do you think the ground temperature changed more than the air temperature? Is that typical on earth?

12. Is there a time when the air and ground were the same temperature? Explain how you know.

Problem Set

1. Create a short written report summarizing your work on the Mars Curiosity Rover Problem. Include your answers to the original problem questions and at least one recommendation for further research on this topic or additional questions you have about the situation.

2. Consider the sky crane descent portion of the landing sequence.

    a. Create a linear function to model the Curiosity Rover's altitude as a function of time. What two points did you choose to create your function?

    b. Compare the slope of your function to the velocity. Should they be equal? Explain why or why not.

    c. Use your linear model to determine the altitude one minute before landing. How does it compare to your earlier estimate? Explain any differences you found.

3. The exponential function $g(t) = 125(0.99)^t$ could be used to model the altitude of the Curiosity Rover during its rapid descent. Do you think this model would be better or worse than the one your group created? Explain your reasoning.

4. For each graph below, identify the increasing and decreasing intervals, the positive and negative intervals, and the intercepts.

    a.
    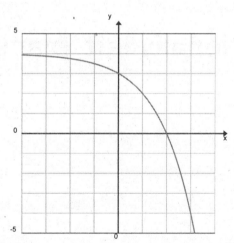

    b.
    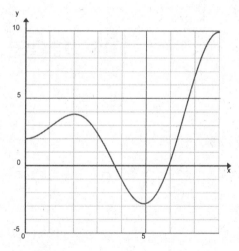

This page intentionally left blank

Lesson 14: Linear and Exponential Models—Comparing Growth Rates

Classwork

Example 1

Linear Functions

a. Sketch points $P_1 = (0,4)$ and $P_2 = (4,12)$. Are there values of m and b such that the graph of the linear function described by $f(x) = mx + b$ contains P_1 and P_2? If so, find those values. If not, explain why they do not exist.

b. Sketch $P_1 = (0,4)$ and $P_2 = (0,-2)$. Are there values of m and b so that the graph of a linear function described by $f(x) = mx + b$ contains P_1 and P_2? If so, find those values. If not, explain why they do not exist.

Exponential Functions

Graphs (c) and (d) are both graphs of an exponential function of the form $g(x) = ab^x$. Rewrite the function $g(x)$ using the values for a and b that are required for the graph shown to be a graph of g.

c. $g(x) =$

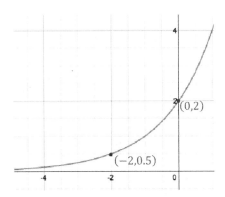

d. $g(x) =$

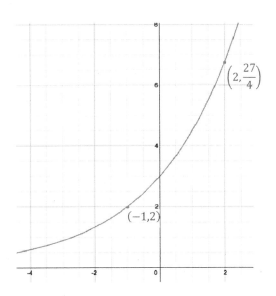

Example 2

A lab researcher records the growth of the population of a yeast colony and finds that the population doubles every hour.

a.   Complete the researcher's table of data:

Hours into study	0	1	2	3	4
Yeast colony population (thousands)	5				

b. What is the exponential function that models the growth of the colony's population?

c. Several hours into the study, the researcher looks at the data and wishes there were more frequent measurements. Knowing that the colony doubles every hour, how can the researcher determine the population in half-hour increments? Explain.

d. Complete the new table that includes half-hour increments.

Hours into study	0	$\frac{1}{2}$	1	$\frac{3}{2}$	2	$\frac{5}{2}$	3
Yeast colony population (thousands)	5						

e. How would the calculation for the data change for time increments of 20 minutes? Explain.

f. Complete the new table that includes 20-minute increments.

Hours into study	0	$\frac{1}{3}$	$\frac{2}{3}$	1	$\frac{4}{3}$	$\frac{5}{3}$	2
Yeast colony population (thousands)	5						

g.   The researcher's lab assistant studies the data recorded and makes the following claim:

Since the population doubles in 1 hour, then half of that growth happens in the first half hour, and the other half of that growth happens in the second half hour. We should be able to find the population at $t = \frac{1}{2}$ by taking the average of the populations at $t = 0$ and $t = 1$.

Is the assistant's reasoning correct? Compare this strategy to your work in parts (c) and (e).

Example 3

A California Population Projection Engineer in 1920 was tasked with finding a model that predicts the state's population growth. He modeled the population growth as a function of time, t years since 1900. Census data shows that the population in 1900, in thousands, was 1,490. In 1920, the population of the state of California was 3,554 thousand. He decided to explore both a linear and an exponential model.

a.   Use the data provided to determine the equation of the linear function that models the population growth from 1900–1920.

b.   Use the data provided and your calculator to determine the equation of the exponential function that models the population growth.

Lesson 14:   Linear and Exponential Models—Comparing Growth Rates

EUREKA MATH

c. Use the two functions to predict the population for the following years:

	Projected Population Based on Linear Function, $f(t)$ (thousands)	Projected Population Based on Exponential Function, $g(t)$ (thousands)	Census Population Data and Intercensal Estimates for California (thousands)
1935			6175
1960			15717
2010			37253

Courtesy U.S. Census Bureau

d. Which function is a better model for the population growth of California in 1935 and in 1960?

e. Does either model closely predict the population for 2010? What phenomenon explains the real population value?

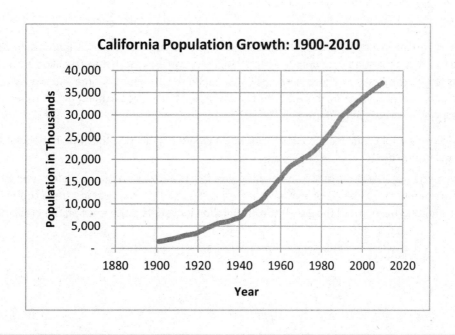

EUREKA
MATH™

Lesson 14: Linear and Exponential Models—Comparing Growth Rates

S.99

This work is derived from Eureka Math ™ and licensed by Great Minds. ©2015 Great Minds. eureka-math.org
ALG1-M3-SE-B1-1.3.0-05.2015

Lesson Summary

- Given a linear function of the form $L(x) = mx + k$ and an exponential function of the form $E(x) = ab^x$ for x a real number and constants m, k, a, and b, consider the sequence given by $L(n)$ and the sequence given by $E(n)$, where $n = 1,2,3,4, \ldots$. Both of these sequences can be written recursively:

$$L(n + 1) = L(n) + m \text{ and } L(0) = k, \text{ and}$$
$$E(n + 1) = E(n) \cdot b \text{ and } E(0) = a.$$

  The first sequence shows that a linear function grows additively by the same summand m over equal length intervals (i.e., the intervals between consecutive integers). The second sequence shows that an exponential function grows multiplicatively by the same factor b over equal-length intervals (i.e., the intervals between consecutive integers).

- An increasing exponential function eventually exceeds any linear function. That is, if $f(x) = ab^x$ is an exponential function with $a > 0$ and $b > 1$, and $g(x) = mx + k$ is a linear function, then there is a real number M such that for all $x > M$, then $f(x) > g(x)$. Sometimes this is not apparent in a graph displayed on a graphing calculator; that is because the graphing window does not show enough of the graphs for us to see the sharp rise of the exponential function in contrast with the linear function.

Problem Set

1. When a ball bounces up and down, the maximum height it reaches decreases with each bounce in a predictable way. Suppose for a particular type of squash ball dropped on a squash court, the maximum height, $h(x)$, after x number of bounces can be represented by $h(x) = 65 \left(\frac{1}{3}\right)^x$.

   a. How many times higher is the height after the first bounce compared to the height after the third bounce?

   b. Graph the points $\big(x, h(x)\big)$ for x-values of 0, 1, 2, 3, 4, and 5.

2. Australia experienced a major pest problem in the early 20th century. The pest? Rabbits. In 1859, 24 rabbits were released by Thomas Austin at Barwon Park. In 1926, there were an estimated 10 billion rabbits in Australia. Needless to say, the Australian government spent a tremendous amount of time and money to get the rabbit problem under control. (To find more on this topic, visit Australia's Department of Environment and Primary Industries website under Agriculture.)

   a. Based only on the information above, write an exponential function that would model Australia's rabbit population growth.

   b. The model you created from the data in the problem is obviously a huge simplification from the actual function of the number of rabbits in any given year from 1859 to 1926. Name at least one complicating factor (about rabbits) that might make the graph of your function look quite different than the graph of the actual function.

Lesson 14:   Linear and Exponential Models—Comparing Growth Rates

3. After graduating from college, Jane has two job offers to consider. Job A is compensated at $100,000 a year but with no hope of ever having an increase in pay. Jane knows a few of her peers are getting that kind of an offer right out of college. Job B is for a social media start-up, which guarantees a mere $10,000 a year. The founder is sure the concept of the company will be the next big thing in social networking and promises a pay increase of 25% at the beginning of each new year.

    a. Which job will have a greater annual salary at the beginning of the fifth year? By approximately how much?

    b. Which job will have a greater annual salary at the beginning of the tenth year? By approximately how much?

    c. Which job will have a greater annual salary at the beginning of the twentieth year? By approximately how much?

    d. If you were in Jane's shoes, which job would you take?

4. The population of a town in 2007 was 15,000 people. The town has gotten its fresh water supply from a nearby lake and river system with the capacity to provide water for up to 30,000 people. Due to its proximity to a big city and a freeway, the town's population has begun to grow more quickly than in the past. The table below shows the population counts for each year from 2007–2012.

    a. Write a function of x that closely matches these data points for x-values of 0, 1, 2, 3, 4, and 5.

Year	Years Past 2007	Population of the town
2007	0	15,000
2008	1	15,600
2009	2	16,224
2010	3	16,873
2011	4	17,548
2012	5	18,250

    b. Assume the function is a good model for the population growth from 2012–2032. At what year during the time frame 2012–2032 will the water supply be inadequate for the population?

This page intentionally left blank

Lesson 15: Piecewise Functions

Classwork

Opening Exercise

For each real number a, the *absolute value of a* is the distance between 0 and a on the number line and is denoted $|a|$.

1.   Solve each one variable equation.

    a. $|x| = 6$ b. $|x - 5| = 4$ c. $2|x + 3| = -10$

2.   Determine at least five solutions for each two-variable equation. Make sure some of the solutions include negative values for either x or y.

    a. $y = |x|$

    b. $y = |x - 5|$

    c. $x = |y|$

Exploratory Challenge 1

For parts (a)–(c) create graphs of the solution set of each two-variable equation from Opening Exercise 2.

    a. $y = |x|$ b. $y = |x - 5|$

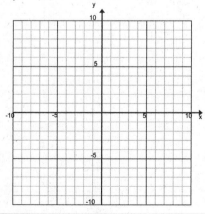
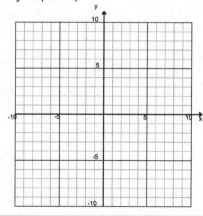

EUREKA MATH™

c. $x = |y|$

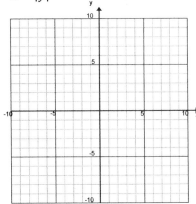

d. Write a brief summary comparing and contrasting the three solution sets and their graphs.

For parts (e)–(j), consider the function $f(x) = |x|$, where x can be any real number.

e. Explain the meaning of the function f in your own words.

f. State the domain and range of this function.

g. Create a graph of the function f. You might start by listing several ordered pairs that represent the corresponding domain and range elements.

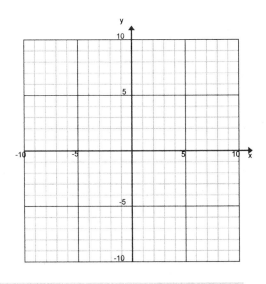

h. How does the graph of the absolute value function compare to the graph of $y = |x|$?

i. Define a function whose graph would be identical to the graph of $y = |x - 5|$.

j. Could you define a function whose graph would be identical to the graph of $x = |y|$? Explain your reasoning.

k. Let $f_1(x) = -x$ for $x < 0$, and let $f_2(x) = x$ for ≥ 0 . Graph the functions f_1 and f_2 on the same Cartesian plane. How does the graph of these two functions compare to the graph in part (g)?

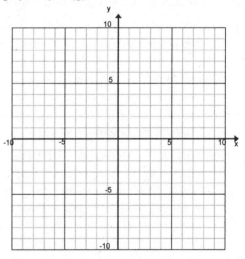

Definition:

The *absolute value function* f is defined by setting $f(x) = |x|$ for all real numbers. Another way to write f is as a piecewise linear function:

$$f(x) = \begin{cases} -x & x < 0 \\ x & x \geq 0 \end{cases}$$

Lesson 15:   Piecewise Functions

S.105

Example 1

Let $g(x) = |x - 5|$. The graph of g is the same as the graph of the equation $y = |x - 5|$ you drew in Exploratory Challenge 1, part (b). Use the redrawn graph below to rewrite the function g as a piecewise function.

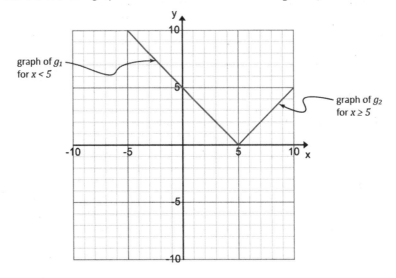

Label the graph of the linear function with negative slope by g_1 and the graph of the linear function with positive slope by g_2, as in the picture above.

Function g_1: The slope of g_1 is -1 (why?), and the y-intercept is 5; therefore, $g_1(x) = -x + 5$.

Function g_2: The slope of g_2 is 1 (why?), and the y-intercept is -5 (why?); therefore, $g_2(x) = x - 5$.

Writing g as a piecewise function is just a matter of collecting all of the different "pieces" and the intervals upon which they are defined:

$$g(x) = \begin{cases} -x + 5 & x < 5 \\ x - 5 & x \geq 5 \end{cases}.$$

Exploratory Challenge 2

The *floor* of a real number x, denoted by $\lfloor x \rfloor$, is the largest integer not greater than x. The *ceiling* of a real number x, denoted by $\lceil x \rceil$, is the smallest integer not less than x. The *sawtooth* number of a positive number is the *fractional part* of the number that is to the right of its floor on the number line. In general, for a real number x, the sawtooth number of x is the value of the expression $x - \lfloor x \rfloor$. Each of these expressions can be thought of as functions with the domain being the set of real numbers.

Lesson 15:   Piecewise Functions

a. Complete the following table to help you understand how these functions assign elements of the domain to elements of the range. The first and second rows have been done for you.

x	$floor(x) = \lfloor x \rfloor$	$ceiling(x) = \lceil x \rceil$	$sawtooth(x) = x - \lfloor x \rfloor$
4.8	4	5	0.8
-1.3	-2	-1	0.7
2.2			
6			
-3			
$-\dfrac{2}{3}$			
π			

b. Create a graph of each function.

$floor(x) = \lfloor x \rfloor$

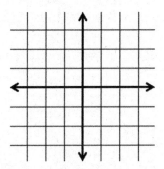

$ceiling(x) = \lceil x \rceil$

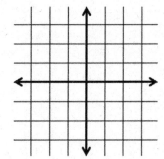

$sawtooth(x) = x - \lfloor x \rfloor$

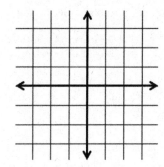

c. For the floor, ceiling, and sawtooth functions, what would be the range values for all real numbers x on the interval $[0,1)$? The interval $(1,2]$? The interval $[-2,-1)$? The interval $[1.5, 2.5]$?

Relevant Vocabulary

PIECEWISE LINEAR FUNCTION: Given a number of nonoverlapping intervals on the real number line, a *(real) piecewise linear function* is a function from the union of the intervals to the set of real numbers such that the function is defined by (possibly different) linear functions on each interval.

ABSOLUTE VALUE FUNCTION: The absolute value of a number x, denoted by $|x|$, is the distance between 0 and x on the number line. The *absolute value function* is the piecewise linear function such that for each real number x, the value of the function is $|x|$.

We often name the absolute value function by saying, "Let $f(x) = |x|$ for all real numbers x."

FLOOR FUNCTION: The *floor* of a real number x, denoted by $\lfloor x \rfloor$, is the largest integer not greater than x. The *floor function* is the piecewise linear function such that for each real number x, the value of the function is $\lfloor x \rfloor$.

We often name the floor function by saying, "Let $f(x) = \lfloor x \rfloor$ for all real numbers x."

CEILING FUNCTION: The *ceiling* of a real number x, denoted by $\lceil x \rceil$, is the smallest integer not less than x. The *ceiling function* is the piecewise linear function such that for each real number x, the value of the function is $\lceil x \rceil$.

We often name the ceiling function by saying, "Let $f(x) = \lceil x \rceil$ for all real numbers x."

SAWTOOTH FUNCTION: The *sawtooth function* is the piecewise linear function such that for each real number x, the value of the function is given by the expression $x - \lfloor x \rfloor$.

The sawtooth function assigns to each positive number the part of the number (the non-integer part) that is to the right of the floor of the number on the number line. That is, if we let $f(x) = x - \lfloor x \rfloor$ for all real numbers x, then

$$f\left(\tfrac{1}{3}\right) = \tfrac{1}{3}, \ f\left(1\tfrac{1}{3}\right) = \tfrac{1}{3}, \ f(1{,}000.02) = 0.02, \ f(-0.3) = 0.7, \text{ etc.}$$

Problem Set

1. Explain why the sawtooth function, $sawtooth(x) = x - \lfloor x \rfloor$ for all real numbers x, takes only the fractional part of a number when the number is positive.

2. Let $g(x) = \lceil x \rceil - \lfloor x \rfloor$, where x can be any real number. In otherwords, g is the difference between the ceiling and floor functions. Express g as a piecewise function.

3. The Heaviside function is defined using the formula below.

$$H(x) = \begin{cases} -1, & x < 0 \\ 0, & x = 0 \\ 1, & x > 0 \end{cases}$$

   Graph this function, and state its domain and range.

4. The following piecewise function is an example of a step function.

$$S(x) = \begin{cases} 3 & -5 \le x < -2 \\ 1 & -2 \le x < 3 \\ 2 & 3 \le x \le 5 \end{cases}$$

   a. Graph this function, and state the domain and range.
   b. Why is this type of function called a step function?

5. Let $f(x) = \dfrac{|x|}{x}$, where x can be any real number except 0.

   a. Why is the number 0 excluded from the domain of f?
   b. What is the range of f?
   c. Create a graph of f.
   d. Express f as a piecewise function.
   e. What is the difference between this function and the Heaviside function?

6. Graph the following piecewise functions for the specified domain.

   a. $f(x) = |x + 3|$ for $-5 \le x \le 3$
   b. $f(x) = |2x|$ for $-3 \le x \le 3$
   c. $f(x) = |2x - 5|$ for $0 \le x \le 5$
   d. $f(x) = |3x + 1|$ for $-2 \le x \le 2$
   e. $f(x) = |x| + x$ for $-5 \le x \le 3$
   f. $f(x) = \begin{cases} x & \text{if } x \le 0 \\ x + 1 & \text{if } x > 0 \end{cases}$
   g. $f(x) = \begin{cases} 2x + 3 & \text{if } x < -1 \\ 3 - x & \text{if } x \ge -1 \end{cases}$

7.   Write a piecewise function for each graph below.

a.

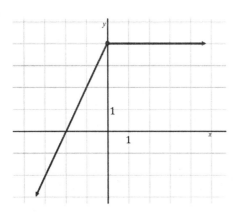

Graph of *b*

b.

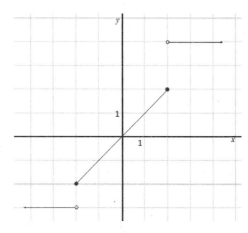

Graph of *p*

c.

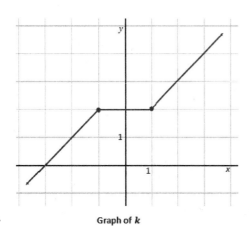

Graph of *k*

d.

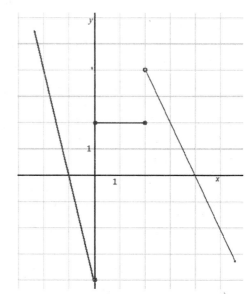

Graph of *h*

EUREKA
MATH™

Lesson 16: Graphs Can Solve Equations Too

Opening Exercise

1. Solve for x in the following equation: $|x + 2| - 3 = 0.5x + 1$.

2. Now, let $f(x) = |x + 2| - 3$ and $g(x) = 0.5x + 1$.
    When does $f(x) = g(x)$?

    a. Graph $y = f(x)$ and $y = g(x)$ on the same set of axes.

    b. When does $f(x) = g(x)$? What is the visual significance
        of the points where $f(x) = g(x)$?

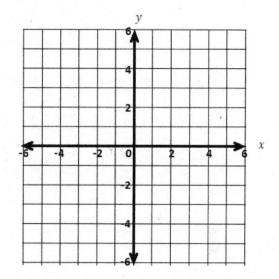

    c. Is each intersection point (x, y) an element of the graph f and an element of the graph of g? In other words,
        do the functions f and g really have the same value when $x = 4$? What about when $x = -4$?

Example 1

Solve this equation by graphing two functions on the same Cartesian plane: $-|x - 3| + 4 = |0.5x| - 5$.

Let $f(x) = -|x - 3| + 4$, and let $g(x) = |0.5x| - 5$, where x can be any real number.

We are looking for values of x at which the functions f and g have the same output value.

Therefore, we set $y = f(x)$ and $y = g(x)$, so we can plot the graphs on the same coordinate plane:

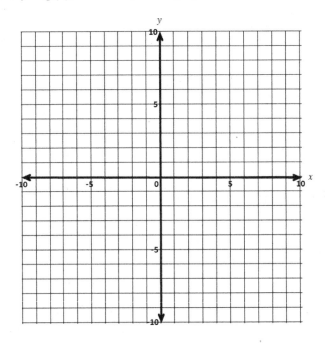

From the graph, we see that the two intersection points are _____ and _____.

The fact that the graphs of the functions meet at these two points means that when x is _____, both $f(x)$ and $g(x)$

are _____, or when x is _____, both $f(x)$ and $g(x)$ are _____.

Thus, the expressions $-|x - 3| + 4$ and $|0.5x| - 5$ are equal when $x =$ _____ or when $x =$ _____.

Therefore, the solution set to the original equation is _____.

Lesson 16:   Graphs Can Solve Equations Too

EUREKA MATH™

Example 2

Solve this equation graphically: $-|x - 3.5| + 4 = -0.25x - 1$.

a.   Write the two functions represented by each side of the equation.

b.   Graph the functions in an appropriate viewing window.

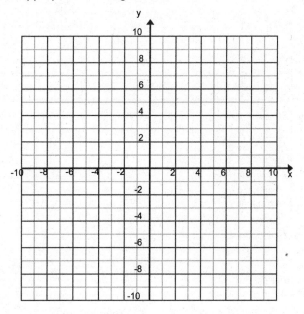

c.   Determine the intersection points of the two functions.

d.   Verify that the x-coordinates of the intersection points are solutions to the equation.

Exercises 1–5

Use graphs to find approximate values of the solution set for each equation. Use technology to support your work. Explain how each of your solutions relates to the graph. Check your solutions using the equation.

1. $3 - 2x = |x - 5|$

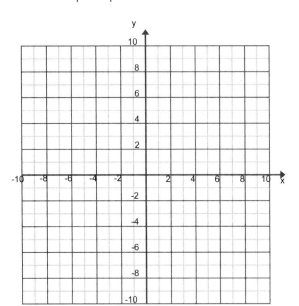

2. $2(1.5)^x = 2 + 1.5x$

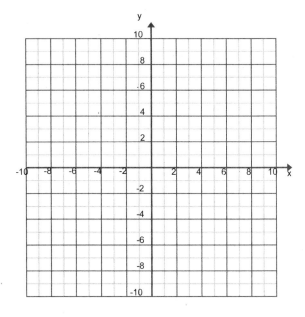

3. The graphs of the functions f and g are shown.

    a. Use the graphs to *approximate* the solution(s) to the equation $f(x) = g(x)$.

    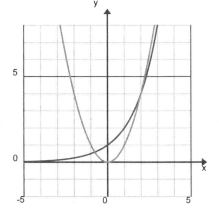

    b. Let $f(x) = x^2$, and let $g(x) = 2^x$. Find *all* solutions to the equation $f(x) = g(x)$. Verify any exact solutions that you determine using the definitions of f and g. Explain how you arrived at your solutions.

4. The graphs of f, a function that involves taking an absolute value, and g, a linear function, are shown to the right. Both functions are defined over all real values for x. Tami concluded that the equation $f(x) = g(x)$ has no solution.

    Do you agree or disagree? Explain your reasoning.

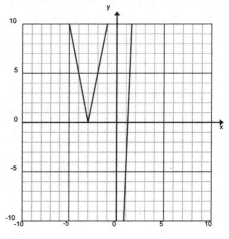

5. The graphs of f (a function that involves taking the absolute value) and g (an exponential function) are shown below. Sharon said the solution set to the equation $f(x) = g(x)$ is exactly $\{-7, 5\}$.

    Do you agree or disagree with Sharon? Explain your reasoning.

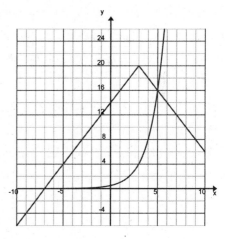

Problem Set

1. Solve the following equations graphically. Verify the solution sets using the original equations.

    a. $|x| = x^2$

    b. $|3x - 4| = 5 - |x - 2|$

2. Find the approximate solution(s) to each of the following equations graphically. Use technology to support your work. Verify the solution sets using the original equations.

    a. $2x - 4 = \sqrt{x + 5}$

    b. $x + 2 = x^3 - 2x - 4$

    c. $0.5x^3 - 4 = 3x + 1$

    d. $6\left(\frac{1}{2}\right)^{5x} = 10 - 6x$

In each problem, the graphs of the functions f and g are shown on the same Cartesian plane. Estimate the solution set to the equation $f(x) = g(x)$. Assume that the graphs of the two functions intersect only at the points shown on the graph.

3.

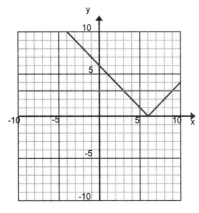

4.

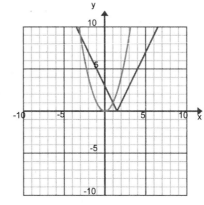

5.

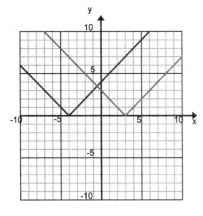

6.

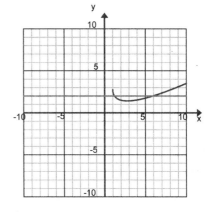

Lesson 16:   Graphs Can Solve Equations Too

EUREKA
MATH™

7. The graph shows Glenn's distance from home as he rode his bicycle to school, which is just down his street. His next-door neighbor Pablo, who lives 100 m closer to the school, leaves his house at the same time as Glenn. He walks at a constant velocity, and they both arrive at school at the same time.

    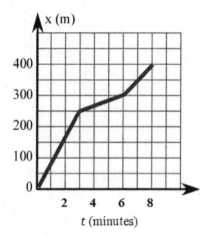

    a. Graph a linear function that represents Pablo's distance from Glenn's home as a function of time.

    b. Estimate when the two boys pass each other.

    c. Write piecewise linear functions to represent each boy's distance, and use them to verify your answer to part (b).

This page intentionally left blank

Lesson 17: Four Interesting Transformations of Functions

Classwork

Exploratory Challenge 1

Let $f(x) = |x|$, $g(x) = f(x) - 3$, and $h(x) = f(x) + 2$ for any real number x.

    a.   Write an explicit formula for $g(x)$ in terms of $|x|$ (i.e., without using $f(x)$ notation).

    b.   Write an explicit formula for $h(x)$ in terms of $|x|$ (i.e., without using $f(x)$ notation).

    c.   Complete the table of values for these functions.

x	$f(x) = |x|$	$g(x) = f(x) - 3$	$h(x) = f(x) + 2$
-3			
-2			
-1			
0			
1			
2			
3			

d.   Graph all three equations: $y = f(x)$, $y = f(x) - 3$, and $y = f(x) + 2$.

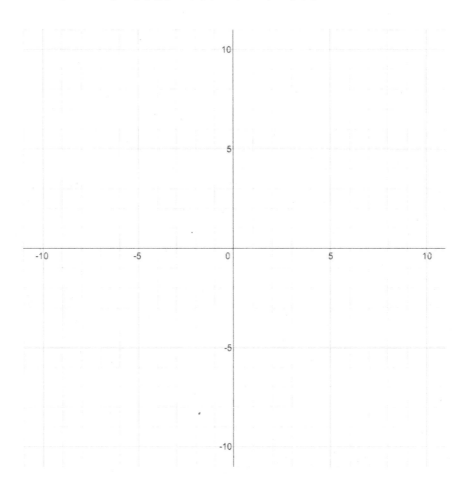

e.   What is the relationship between the graph of $y = f(x)$ and the graph of $y = f(x) + k$?

f.   How do the values of g and h relate to the values of f?

Lesson 17:   Four Interesting Transformations of Functions

This work is derived from Eureka Math ™ and licensed by Great Minds. ©2015 Great Minds. eureka-math.org
ALG1-M3-SE-B1-1.3.0-05.2015

EUREKA
MATH™

Exploratory Challenge 2

Let $f(x) = |x|$, $g(x) = 2f(x)$, and $h(x) = \frac{1}{2}f(x)$ for any real number x.

a. Write a formula for $g(x)$ in terms of $|x|$ (i.e., without using $f(x)$ notation).

b. Write a formula for $h(x)$ in terms of $|x|$ (i.e., without using $f(x)$ notation).

c. Complete the table of values for these functions.

x	$f(x) = |x|$	$g(x) = 2f(x)$	$h(x) = \frac{1}{2}f(x)$
-3			
-2			
-1			
0			
1			
2			
3			

EUREKA MATH™

d. Graph all three equations: $y = f(x)$, $y = 2f(x)$, and $y = \frac{1}{2}f(x)$.

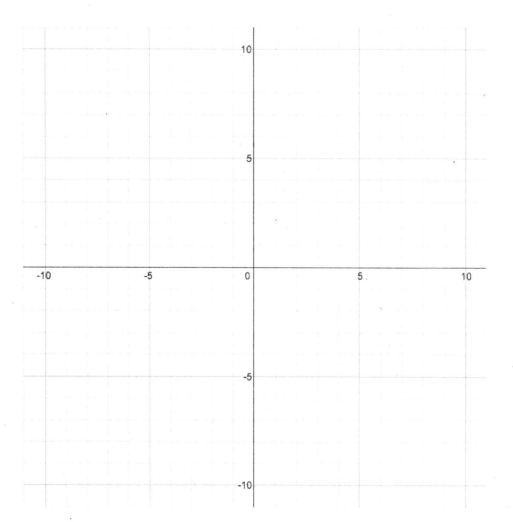

Given $f(x) = |x|$, let $p(x) = -|x|$, $q(x) = -2f(x)$, and $r(x) = -\frac{1}{2}f(x)$ for any real number x.

e. Write the formula for $q(x)$ in terms of $|x|$ (i.e., without using $f(x)$ notation).

f. Write the formula for $r(x)$ in terms of $|x|$ (i.e., without using $f(x)$ notation).

Lesson 17:   Four Interesting Transformations of Functions

EUREKA
MATH™

g. Complete the table of values for the functions $p(x) = -|x|$, $q(x) = -2f(x)$, and $r(x) = -\frac{1}{2}f(x)$.

x	$p(x) = -|x|$	$q(x) = -2f(x)$	$r(x) = -\frac{1}{2}f(x)$
-3			
-2			
-1			
0			
1			
2			
3			

h. Graph all three functions on the same graph that was created in part (d). Label the graphs as $y = p(x)$, $y = q(x)$, and $y = r(x)$.

i. How is the graph of $y = f(x)$ related to the graph of $y = kf(x)$ when $k > 1$?

j. How is the graph of $y = f(x)$ related to the graph of $y = kf(x)$ when $0 < k < 1$?

k. How do the values of functions p, q, and r relate to the values of functions f, g, and h, respectively? What transformation of the graphs of f, g, and h represents this relationship?

Exercise

Make up your own function f by drawing the graph of it on the Cartesian plane below. Label it as the graph of the equation $y = f(x)$. If $b(x) = f(x) - 4$ and $c(x) = \frac{1}{4} f(x)$ for every real number x, graph the equations $y = b(x)$ and $y = c(x)$ on the same Cartesian plane.

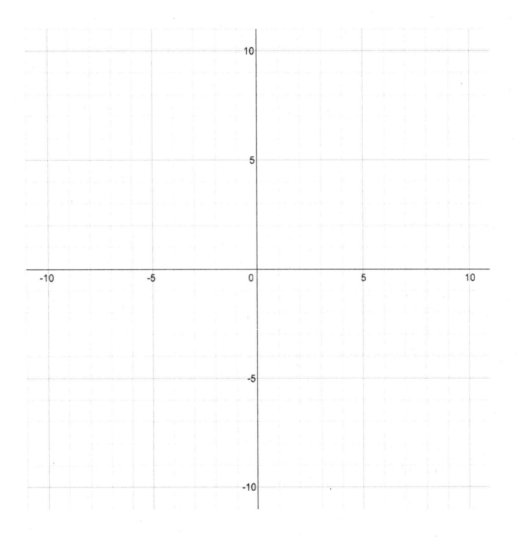

EUREKA
MATH™

Problem Set

Let $f(x) = |x|$ for every real number x. The graph of $y = f(x)$ is shown below. Describe how the graph for each function below is a transformation of the graph of $y = f(x)$. Then, use this same set of axes to graph each function for Problems 1–5. Be sure to label each function on your graph (by $y = a(x)$, $y = b(x)$, etc.).

1. $a(x) = |x| + \dfrac{3}{2}$

2. $b(x) = -|x|$

3. $c(x) = 2|x|$

4. $d(x) = \dfrac{1}{3}|x|$

5. $e(x) = |x| - 3$

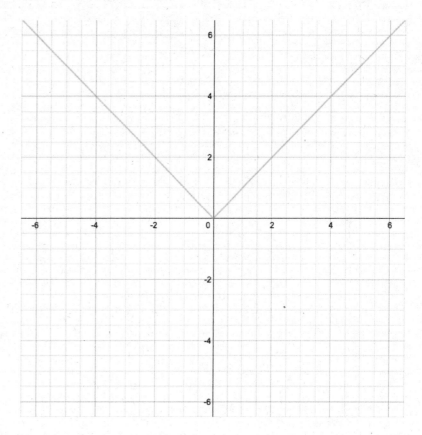

6. Let $r(x) = |x|$ and $t(x) = -2|x| + 1$ for every real number x. The graph of $y = r(x)$ is shown below. Complete the table below to generate output values for the function t, and then graph the equation $y = t(x)$ on the same set of axes as the graph of $y = r(x)$.

x	$r(x) = |x|$	$t(x) = -2|x| + 1$
-2		
-1		
0		
1		
2		

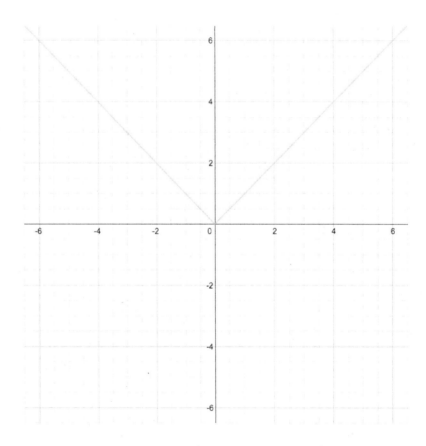

Lesson 17:   Four Interesting Transformations of Functions

This work is derived from Eureka Math ™ and licensed by Great Minds. ©2015 Great Minds. eureka-math.org
ALG1-M3-SE-B1-1.3.0-05.2015

7. Let $f(x) = |x|$ for every real number x. Let m and n be functions found by transforming the graph of $y = f(x)$. Use the graphs of $y = f(x)$, $y = m(x)$, and $y = n(x)$ below to write the functions m and n in terms of the function f. (Hint: What is the k?)

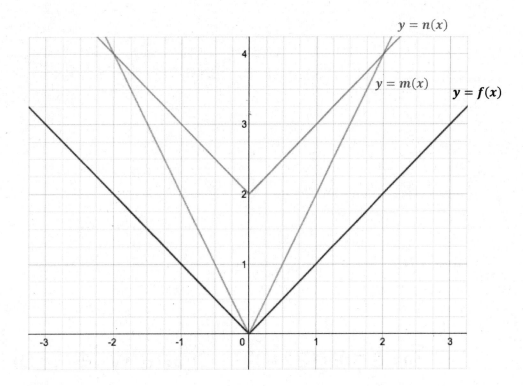

This page intentionally left blank

Lesson 18: Four Interesting Transformations of Functions

Classwork

Example

Let $f(x) = |x|$, $g(x) = f(x - 3)$, and $h(x) = f(x + 2)$, where x can be any real number.

   a.   Write the formula for $g(x)$ in terms of $|x|$ (i.e., without using $f(x)$ notation).

   b.   Write the formula for $h(x)$ in terms of $|x|$ (i.e., without using $f(x)$ notation).

   c.   Complete the table of values for these functions.

x	$f(x) = |x|$	$g(x) =$	$h(x) =$
-3			
-2			
-1			
0			
1			
2			
3			

d.   Graph all three equations: $y = f(x)$, $y = f(x - 3)$, and $y = f(x + 2)$.

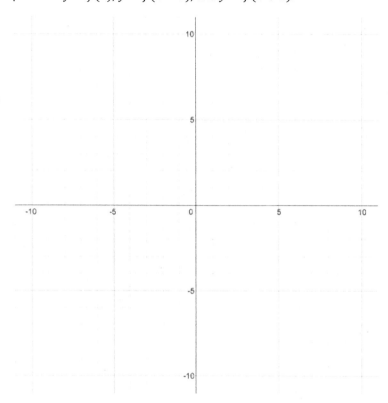

e.   How does the graph of $y = f(x)$ relate to the graph of $y = f(x - 3)$?

f.   How does the graph of $y = f(x)$ relate to the graph of $y = f(x + 2)$?

g.   How do the graphs of $y = |x| - 3$ and $y = |x - 3|$ relate differently to the graph of $y = |x|$?

h.   How do the values of g and h relate to the values of f?

EUREKA
MATH™

Exercises

1. Karla and Isamar are disagreeing over which way the graph of the function $g(x) = |x + 3|$ is translated relative to the graph of $f(x) = |x|$. Karla believes the graph of g is "to the right" of the graph of f; Isamar believes the graph is "to the left." Who is correct? Use the coordinates of the vertex of f and g to support your explanation.

2. Let $f(x) = |x|$, where x can be any real number. Write a formula for the function whose graph is the transformation of the graph of f given by the instructions below.

    a. A translation right 5 units

    b. A translation down 3 units

    c. A vertical scaling (a vertical stretch) with scale factor of 5

    d. A translation left 4 units

    e. A vertical scaling (a vertical shrink) with scale factor of $\dfrac{1}{3}$

3. Write the formula for the function depicted by the graph.

a. $y =$

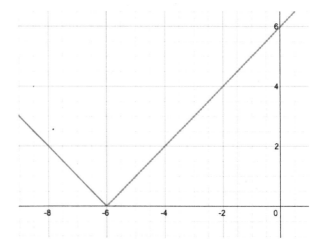

b. $y =$

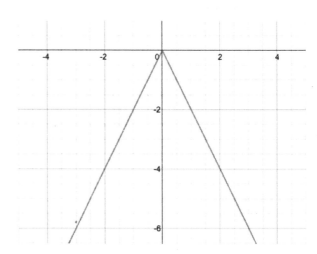

c. $y =$

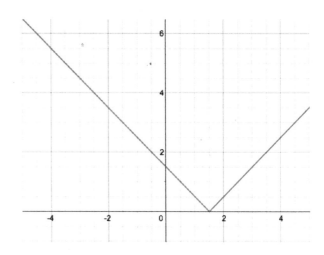

EUREKA
MATH™

d. $y =$

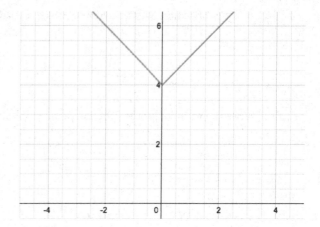

e. $y =$

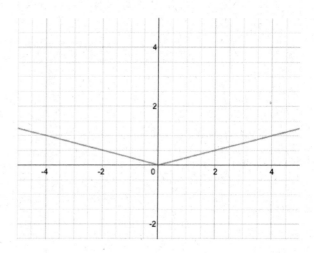

4.   Let $f(x) = |x|$, where x can be any real number. Write a formula for the function whose graph is the described transformation of the graph of f.

a.   A translation 2 units left and 4 units down

b.   A translation 2.5 units right and 1 unit up

c.   A vertical scaling with scale factor $\dfrac{1}{2}$ and then a translation 3 units right

EUREKA
MATH™

    d.   A translation 5 units right and a vertical scaling by reflecting across the x-axis with vertical scale factor -2

5.   Write the formula for the function depicted by the graph.

    a. $y =$

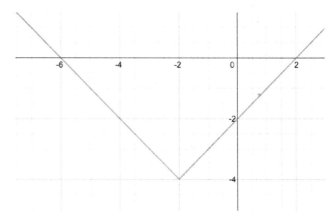

    b. $y =$

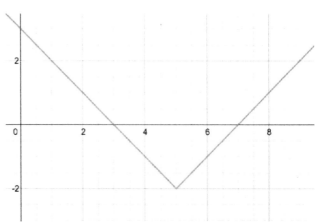

    c. $y =$

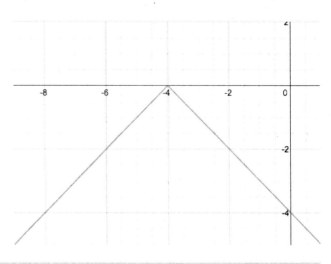

Lesson 18:   Four Interesting Transformations of Functions

EUREKA
MATH™

d. $y =$

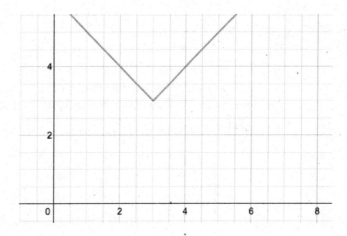

Problem Set

1. Working with quadratic functions:

   a. The vertex of the quadratic function $f(x) = x^2$ is at $(0,0)$, which is the minimum for the graph of f. Based on your work in this lesson, to where do you predict the vertex will be translated for the graphs of $g(x) = (x - 2)^2$ and $h(x) = (x + 3)^2$?

   b. Complete the table of values, and then graph all three functions.

x	$f(x) = x^2$	$g(x) = (x - 2)^2$	$h(x) = (x + 3)^2$
−3			
−2			
−1			
0			
1			
2			
3			

2. Let $f(x) = |x - 4|$ for every real number x. The graph of the equation $y = f(x)$ is provided on the Cartesian plane below. Transformations of the graph of $y = f(x)$ are described below. After each description, write the equation for the transformed graph. Then, sketch the graph of the equation you write for part (d).

   a. Translate the graph left 6 units and down 2 units.

   b. Reflect the resulting graph from part (a) across the x-axis.

   c. Scale the resulting graph from part (b) vertically by a scale factor of $\frac{1}{2}$.

   d. Translate the resulting graph from part (c) right 3 units and up 2 units. Graph the resulting equation.

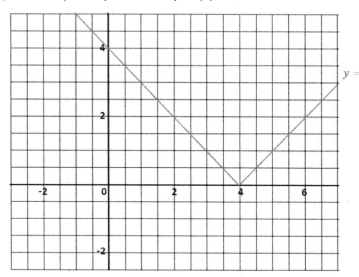

$y = |x - 4|$

Lesson 18:   Four Interesting Transformations of Functions

EUREKA
MATH™

3. Let $f(x) = |x|$ for all real numbers x. Write the formula for the function represented by the described transformation of the graph of $y = f(x)$.

    a. First, a vertical stretch with scale factor $\frac{1}{3}$ is performed, then a translation right 3 units, and finally a translation down 1 unit.

    b. First, a vertical stretch with scale factor 3 is performed, then a reflection over the x-axis, then a translation left 4 units, and finally a translation up 5 units.

    c. First, a reflection across the x-axis is performed, then a translation left 4 units, then a translation up 5 units, and finally a vertical stretch with scale factor 3.

    d. Compare your answers to parts (b) and (c). Why are they different?

4. Write the formula for the function depicted by each graph.

    a. $a(x) =$

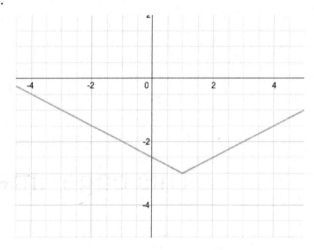

    b. $b(x) =$

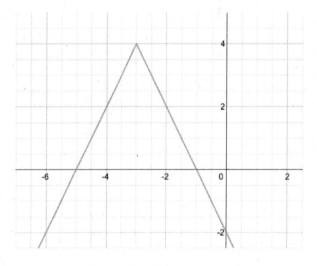

This page intentionally left blank

Lesson 19: Four Interesting Transformations of Functions

Classwork

Exploratory Challenge 1

Let $f(x) = x^2$ and $g(x) = f(2x)$, where x can be any real number.

a. Write the formula for g in terms of x^2 (i.e., without using $f(x)$ notation).

b. Complete the table of values for these functions.

x	$f(x) = x^2$	$g(x) = f(2x)$
-3		
-2		
-1		
0		
1		
2		
3		

c. Graph both equations: $y = f(x)$ and $y = f(2x)$.

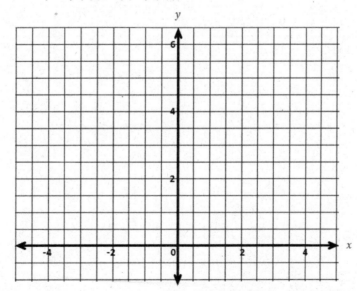

d. How does the graph of $y = g(x)$ relate to the graph of $y = f(x)$?

e. How are the values of f related to the values of g?

Exploratory Challenge 2

Let $f(x) = x^2$ and $h(x) = f\left(\frac{1}{2}x\right)$, where x can be any real number.

a. Rewrite the formula for h in terms of x^2 (i.e., without using $f(x)$ notation).

b. Complete the table of values for these functions.

x	$f(x) = x^2$	$h(x) = f\left(\frac{1}{2}x\right)$
-3		
-2		
-1		
0		
1		
2		
3		

EUREKA MATH™

c.   Graph both equations: $y = f(x)$ and $y = f\left(\frac{1}{2}x\right)$.

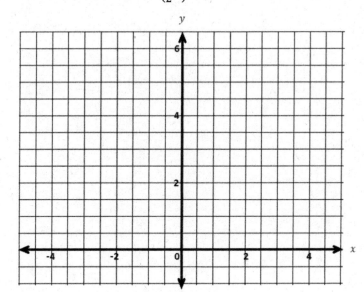

d.   How does the graph of $y = f(x)$ relate to the graph of $y = h(x)$?

e.   How are the values of f related to the values of h?

Exercise

Complete the table of values for the given functions.

a.

x	$f(x) = 2^x$	$g(x) = 2^{(2x)}$	$h(x) = 2^{(-x)}$
-2			
-1			
0			
1			
2			

b. Label each of the graphs with the appropriate functions from the table.

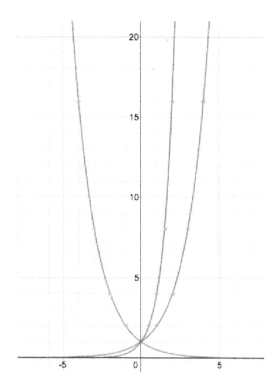

c. Describe the transformation that takes the graph of $y = f(x)$ to the graph of $y = g(x)$.

d. Consider $y = f(x)$ and $y = h(x)$. What does negating the input do to the graph of f?

e. Write the formula of an exponential function whose graph would be a horizontal stretch relative to the graph of g.

EUREKA
MATH™

Exploratory Challenge 3

a. Look at the graph of $y = f(x)$ for the function $f(x) = x^2$ in Exploratory Challenge 1 again. Would we see a difference in the graph of $y = g(x)$ if -2 were used as the scale factor instead of 2? If so, describe the difference. If not, explain why not.

b. A reflection across the y-axis takes the graph of $y = f(x)$ for the function $f(x) = x^2$ back to itself. Such a transformation is called a *reflection symmetry*. What is the equation for the graph of the reflection symmetry of the graph of $y = f(x)$?

c. Deriving the answer to the following question is fairly sophisticated; do this only if you have time. In Lessons 17 and 18, we used the function $f(x) = |x|$ to examine the graphical effects of transformations of a function. In this lesson, we use the function $f(x) = x^2$ to examine the graphical effects of transformations of a function. Based on the observations you made while graphing, why would using $f(x) = x^2$ be a better option than using the function $f(x) = |x|$?

Problem Set

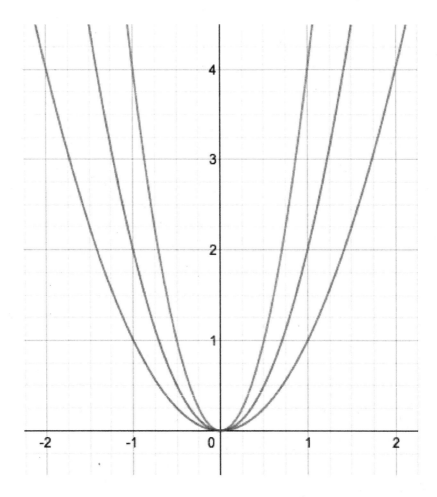

Let $f(x) = x^2$, $g(x) = 2x^2$, and $h(x) = (2x)^2$, where x can be any real number. The graphs above are of the functions $y = f(x)$, $y = g(x)$, and $y = h(x)$.

1. Label each graph with the appropriate equation.

2. Describe the transformation that takes the graph of $y = f(x)$ to the graph of $y = g(x)$. Use coordinates to illustrate an example of the correspondence.

3. Describe the transformation that takes the graph of $y = f(x)$ to the graph of $y = h(x)$. Use coordinates to illustrate an example of the correspondence.

EUREKA
MATH™

Lesson 20: Four Interesting Transformations of Functions

Classwork

Opening Exercise

Fill in the blanks of the table with the appropriate heading or descriptive information.

Graph of $y = f(x)$		Vertical			Horizontal	
Translate	$y = f(x) + k$	$k > 0$	Translate up by $\|k\|$ units		$k > 0$	Translate right by $\|k\|$ units
			Translate down by $\|k\|$ units		$k < 0$	
Scale by scale factor k		$k > 1$				Horizontal stretch by a factor of $\|k\|$
		$0 < k < 1$	Vertical shrink by a factor of $\|k\|$	$y = f\left(\dfrac{1}{k}x\right)$	$0 < k < 1$	
			Vertical shrink by a factor of $\|k\|$ and reflection over x-axis		$-1 < k < 0$	Horizontal shrink by a factor of $\|k\|$ and reflection across y-axis
		$k < -1$			$k < -1$	Horizontal stretch by a factor of $\|k\|$ and reflection over y-axis

Exploratory Challenge 1

A transformation of the absolute value function $f(x) = |x - 3|$ is rewritten here as a piecewise function. Describe in words how to graph this piecewise function.

$$f(x) = \begin{cases} -x + 3, & x < 3 \\ x - 3, & x \geq 3 \end{cases}$$

Exercises 1–2

1. Describe how to graph the following piecewise function. Then, graph $y = f(x)$ below.

$$f(x) = \begin{cases} -3x - 3, & x \leq -2 \\ 0.5x + 4, & -2 < x < 2 \\ -2x + 9, & x \geq 2 \end{cases}$$

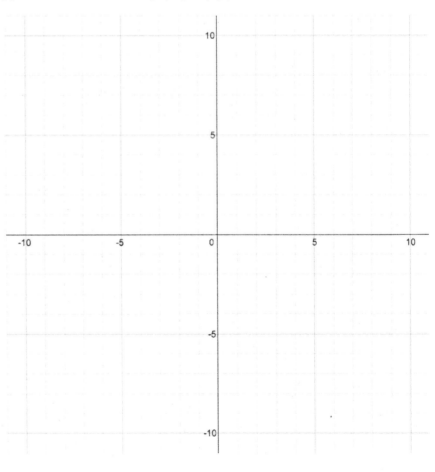

Lesson 20:   Four Interesting Transformations of Functions

EUREKA
MATH™

2. Using the graph of f below, write a formula for f as a piecewise function.

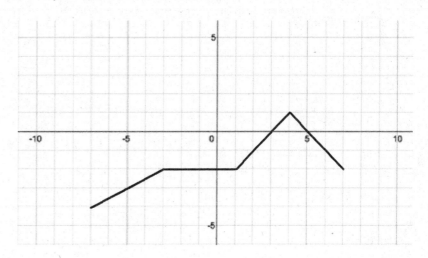

Exploratory Challenge 2

The graph $y = f(x)$ of a piecewise function f is shown. The domain of f is $-5 \le x \le 5$, and the range is $-1 \le y \le 3$.

a. Mark and identify four strategic points helpful in sketching the graph of $y = f(x)$.

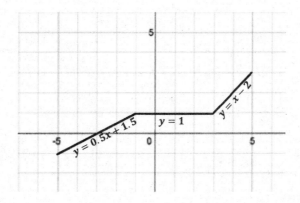

b.   Sketch the graph of $y = 2f(x)$, and state the domain and range of the transformed function. How can you use part (a) to help sketch the graph of $y = 2f(x)$?

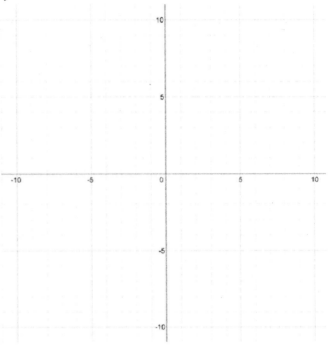

c.   A horizontal scaling with scale factor $\frac{1}{2}$ of the graph of $y = f(x)$ is the graph of $y = f(2x)$. Sketch the graph of $y = f(2x)$, and state the domain and range. How can you use the points identified in part (a) to help sketch $y = f(2x)$?

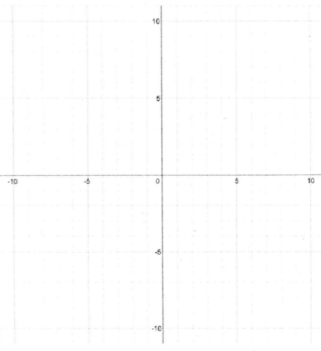

Lesson 20:   Four Interesting Transformations of Functions

EUREKA
MATH™

Exercises 3–4

3. How does the range of f in Exploratory Challenge 2 compare to the range of a transformed function g, where $g(x) = kf(x)$, when $k > 1$?

4. How does the domain of f in Exploratory Challenge 2 compare to the domain of a transformed function g, where $g(x) = f\left(\frac{1}{k}x\right)$, when $0 < k < 1$? (Hint: How does a graph shrink when it is horizontally scaled by a factor k?)

EUREKA
MATH™

Problem Set

1. Suppose the graph of f is given. Write an equation for each of the following graphs after the graph of f has been transformed as described. Note that the transformations are not cumulative.

    a. Translate 5 units upward.

    b. Translate 3 units downward.

    c. Translate 2 units right.

    d. Translate 4 units left.

    e. Reflect about the x-axis.

    f. Reflect about the y-axis.

    g. Stretch vertically by a factor of 2.

    h. Shrink vertically by a factor of $\dfrac{1}{3}$.

    i. Shrink horizontally by a factor of $\dfrac{1}{3}$.

    j. Stretch horizontally by a factor of 2.

2. Explain how the graphs of the equations below are related to the graph of $y = f(x)$.

    a. $y = 5f(x)$

    b. $y = f(x - 4)$

    c. $y = -2f(x)$

    d. $y = f(3x)$

    e. $y = 2f(x) - 5$

This work is derived from Eureka Math ™ and licensed by Great Minds. ©2015 Great Minds. eureka-math.org
ALG1-M3-SE-B1-1.3.0-05.2015

EUREKA
MATH™

3. The graph of the equation $y = f(x)$ is provided below. For each of the following transformations of the graph, write a formula (in terms of f) for the function that is represented by the transformation of the graph of $y = f(x)$. Then, draw the transformed graph of the function on the same set of axes as the graph of $y = f(x)$.

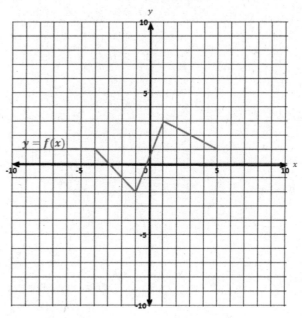

a. A translation 3 units left and 2 units up

b. A vertical stretch by a scale factor of 3

c. A horizontal shrink by a scale factor of $\frac{1}{2}$

4. Reexamine your work on Exploratory Challenge 2 and Exercises 3 and 4 from this lesson. Parts (b) and (c) of Exploratory Challenge 2 asked how the equations $y = 2f(x)$ and $y = f(2x)$ could be graphed with the help of the strategic points found in part (a). In this problem, we investigate whether it is possible to determine the graphs of $y = 2f(x)$ and $y = f(2x)$ by working with the piecewise linear function f directly.

a. Write the function f in Exploratory Challenge 2 as a piecewise linear function.

b. Let $g(x) = 2f(x)$. Use the graph you sketched in Exploratory Challenge 2, part (b) of $y = 2f(x)$ to write the formula for the function g as a piecewise linear function.

c. Let $h(x) = f(2x)$. Use the graph you sketched in Exploratory Challenge 2, part (c) of $y = f(2x)$ to write the formula for the function h as a piecewise linear function.

d. Compare the piecewise linear functions g and h to the piecewise linear function f. Did the expressions defining each piece change? If so, how? Did the domains of each piece change? If so how?

EUREKA MATH™

This page intentionally left blank

Lesson 21: Comparing Linear and Exponential Models Again

Classwork

Opening Exercise

	Linear Model	**Exponential Model**
General Form		
Meaning of Parameters a and b		
Example		
Rule for Finding $f(x + 1)$ from $f(x)$		
Table		
Graph		
Story Problem Example		

Exercises

1. For each table below, assume the function f is defined for all real numbers. Calculate $\Delta f = f(x + 1) - f(x)$ in the last column in the tables below, and show your work. (The symbol Δ in this context means *change in*.) What do you notice about Δf? Could the function be linear or exponential? Write a linear or an exponential function formula that generates the same input–output pairs as given in the table.

x	$f(x)$	$\Delta f = f(x + 1) - f(x)$
1	-3	
2	1	
3	5	
4	9	
5	13	

x	$f(x)$	$\Delta f = f(x + 1) - f(x)$
0	2	
1	6	
2	18	
3	54	
4	162	

EUREKA
MATH™

2. Terence looked down the second column of the table and noticed that $\dfrac{3}{1} = \dfrac{9}{3} = \dfrac{27}{9} = \dfrac{81}{27}$. Because of his observation, he claimed that the input-output pairs in this table could be modeled with an exponential function. Explain why Terence is correct or incorrect. If he is correct, write a formula for the exponential function that generates the input-output pairs given in the table. If he is incorrect, determine and write a formula for a function that generates the input-output pairs given in the table.

x	$T(x)$
0	1
1	3
4	9
13	27
40	81

3. A river has an initial minnow population of 40,000 that is growing at 5% per year. Due to environmental conditions, the amount of algae that minnows use for food is decreasing, supporting 1,000 fewer minnows each year. Currently, there is enough algae to support 50,000 minnows. Is the minnow population increasing linearly or exponentially? Is the amount of algae decreasing at a linear or an exponential rate? In what year will the minnow population exceed the amount of algae available?

4. Using a calculator, Joanna made the following table and then made the following conjecture: $3x$ is always greater than $(1.02)^x$. Is Joanna correct? Explain.

x	$(1.02)^x$	$3x$
1	1.02	3
2	1.0404	6
3	1.0612	9
4	1.0824	12
5	1.1041	15

EUREKA
MATH™

Lesson Summary

- Suppose that the input-output pairs of a bivariate data set have the following property: For every two inputs that are a given difference apart, the difference in their corresponding outputs is constant. Then, an appropriate model for that data set could be a linear function.

- Suppose that the input-output pairs of a bivariate data set have the following property: For every two inputs that are a given difference apart, the quotient of their corresponding outputs is constant. Then, an appropriate model for that data set could be an exponential function.

- An increasing exponential function will eventually exceed any linear function. That is, if $f(x) = ab^x$ is an exponential function with $a > 0$ and $b > 1$, and $g(x) = mx + k$ is any linear function, then there is a real number M such that for all $x > M$, then $f(x) > g(x)$. Sometimes this is not apparent in a graph displayed on a graphing calculator; that is because the graphing window does not show enough of the graph to show the sharp rise of the exponential function in contrast with the linear function.

Problem Set

For each table in Problems 1–6, classify the data as describing a linear relationship, an exponential growth relationship, an exponential decay relationship, or neither. If the relationship is linear, calculate the constant rate of change (slope), and write a formula for the linear function that models the data. If the function is exponential, calculate the common quotient for input values that are distance one apart, and write the formula for the exponential function that models the data. For each linear or exponential function found, graph the equation $y = f(x)$.

1.

x	$f(x)$
1	$\dfrac{1}{2}$
2	$\dfrac{1}{4}$
3	$\dfrac{1}{8}$
4	$\dfrac{1}{16}$
5	$\dfrac{1}{32}$

EUREKA
MATH™

2.

x	$f(x)$
1	1.4
2	2.5
3	3.6
4	4.7
5	5.8

3.

x	$f(x)$
1	−1
2	0
3	2
4	5
5	9

4.

x	$f(x)$
1	20
2	40
3	80
4	160
5	320

5.

x	$f(x)$
1	-5
2	-12
3	-19
4	-26
5	-33

6.

x	$f(x)$
1	$\dfrac{1}{2}$
2	$\dfrac{1}{3}$
3	$\dfrac{1}{4}$
4	$\dfrac{1}{5}$
5	$\dfrac{1}{6}$

7. Here is a variation on a classic riddle: Jayden has a dog-walking business. He has two plans. Plan 1 includes walking a dog once a day for a rate of $5 per day. Plan 2 also includes one walk a day but charges 1 cent for 1 day, 2 cents for 2 days, 4 cents for 3 days, and 8 cents for 4 days and continues to double for each additional day. Mrs. Maroney needs Jayden to walk her dog every day for two weeks. Which plan should she choose? Show the work to justify your answer.

8. Tim deposits money in a certificate of deposit account. The balance (in dollars) in his account t years after making the deposit is given by $T(t) = 1000(1.06)^t$ for $t \geq 0$.

   a. Explain, in terms of the structure of the expression used to define $T(t)$, why Tim's balance can never be $999.

   b. By what percent does the value of $T(t)$ grow each year? Explain by writing a recursive formula for the sequence $T(1)$, $T(2)$, $T(3)$, etc.

   c. By what percentages does the value of $T(t)$ grow every two years? (Hint: Use your recursive formula to write $T(n + 2)$ in terms of $T(n)$.)

9. Your mathematics teacher asks you to sketch a graph of the exponential function $f(x) = \left(\frac{3}{2}\right)^x$ for x, a number between 0 and 40 inclusively, using a scale of 10 units to one inch for both the x- and y-axes.

   a. What are the dimensions (in feet) of the roll of paper needed to sketch this graph?

   b. How many more feet of paper would you need to add to the roll in order to graph the function on the interval $0 \le x \le 41$?

   c. Find an m so that the linear function $g(x) = mx + 2$ is greater than $f(x)$ for all x such that $0 \le x \le 40$, but $f(41) > g(41)$.

EUREKA
MATH™

Lesson 22: Modeling an Invasive Species Population

Classwork

Mathematical Modeling Exercise

The lionfish is a fish that is native to the western Pacific Ocean. The lionfish began appearing in the western Atlantic Ocean in 1985. This is probably because people bought them as pets and then dumped them in waterways leading to the ocean. Because it has no natural predators in this area, the number of lionfish grew very quickly and now has large populations throughout the Caribbean as well as along the eastern coastline of the United States and the Gulf of Mexico. Lionfish have recently been spotted as far north as New York and Rhode Island.

The table below shows the number of new sightings by year reported to NAS (Nonindigenous Aquatic Species), which is a branch of the U.S. Geological Survey Department.

Year	Number of New Sightings	Total Number of Sightings
1985	1	
1992	1	
1995	3	
1996	1	
2000	6	
2001	25	
2002	50	
2003	45	
2004	57	
2005	43	
2006	51	
2007	186	
2008	173	
2009	667	
2010	1,342	

1. Complete the table by recording the total number of sightings for each year.

2. Examine the total number of sightings data. Which model appears to be a better fit for the data—linear or exponential? Explain your reasoning.

3. Make a scatter plot of the year versus the total number of sightings.

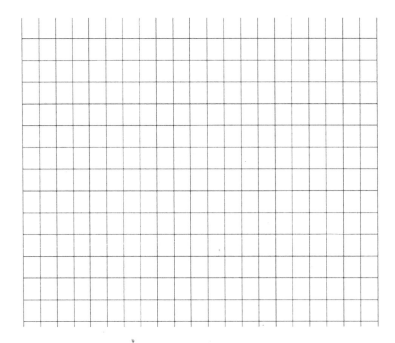

4. Based on the scatter plot, either revise your answer from Exercise 2 or explain how the scatter plot supports your answer from Exercise 2.

5. On the scatter plot, draw a smooth curve that best fits the data.

Lesson 22:   Modeling an Invasive Species Population

**EUREKA
MATH™**

6. From your table, calculate the average rate of change in the total number of sightings for each of the following time intervals.

   a.   1995–2000

   b.   2000–2005

   c.   2005–2010

7. How do the average rates of change help to support your argument of whether a linear or an exponential model is better suited for the data?

8. Use the regression feature of a graphing calculator to find an equation that models the number of lionfish sightings each year.

9. Use your model to predict the total number of lionfish sightings by the end of 2013.

10. The actual number of sightings as of July 2013 was 3,776. Does it seem that your model produced an accurate prediction? Explain.

Problem Set

Another Invasive Species Problem: Kudzu

Kudzu, a perennial vine native to Southeast Asia, now covers a large area of the southern United States. Kudzu was promoted as a forage crop and an ornamental plant when it was introduced to the U.S. at the Philadelphia Centennial Exposition in 1876. Many southern farmers were encouraged to plant kudzu for erosion control from the mid-1930s to the mid-1950s. In 1953, kudzu was removed from the U.S. Department of Agriculture's list of permissible cover plants due to its recognition as an invasive species.

Look up information about kudzu in the U.S. on Wikipedia, and write a short (1- to 2-page) report on the growth of kudzu since its introduction. In your report, choose a function (linear or exponential) to model and graph the growth of kudzu (in hectares) in the U.S. per year over the past half century or so. Remember to cite your sources!

Lesson 23: Newton's Law of Cooling

Classwork

Opening Exercise

A detective is called to the scene of a crime where a dead body has just been found. He arrives at the scene and measures the temperature of the dead body at 9:30 p.m. After investigating the scene, he declares that the person died 10 hours prior, at approximately 11:30 a.m. A crime scene investigator arrives a little later and declares that the detective is wrong. She says that the person died at approximately 6:00 a.m., 15.5 hours prior to the measurement of the body temperature. She claims she can prove it by using Newton's law of cooling:

$$T(t) = T_a + (T_0 - T_a) \cdot 2.718^{-kt},$$

where:

$T(t)$ is the temperature of the object after a time of t hours has elapsed,

T_a is the ambient temperature (the temperature of the surroundings), assumed to be constant, not impacted by the cooling process,

T_0 is the initial temperature of the object, and

k is the decay constant.

Using the data collected at the scene, decide who is correct: the detective or the crime scene investigator.

$T_a = 68°F$ (the temperature of the room)

$T_0 = 98.6°F$ (the initial temperature of the body)

$k = 0.1335$ (13.35% per hour—calculated by the investigator from the data collected)

The temperature of the body at 9:30 p.m. is 72°F.

Mathematical Modeling Exercise

Two cups of coffee are poured from the same pot. The initial temperature of the coffee is $180°F$, and k is 0.2337 (for time in minutes).

1. Suppose both cups are poured at the same time. Cup 1 is left sitting in the room that is $75°F$, and Cup 2 is taken outside where it is $42°F$.

   a. Use Newton's law of cooling to write equations for the temperature of each cup of coffee after t minutes has elapsed.

   b. Graph and label both on the same coordinate plane, and compare and contrast the two graphs.

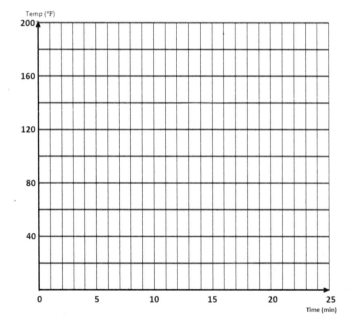

EUREKA
MATH™

c. Coffee is safe to drink when its temperature is below 140°F. Estimate how much time elapses before each cup is safe to drink.

2. Suppose both cups are poured at the same time, and both are left sitting in the room that is 75°F. But this time, milk is immediately poured into Cup 2, cooling it to an initial temperature of 162°F.

a. Use Newton's law of cooling to write equations for the temperature of each cup of coffee after t minutes has elapsed.

b. Graph and label both on the same coordinate plane, and compare and contrast the two graphs.

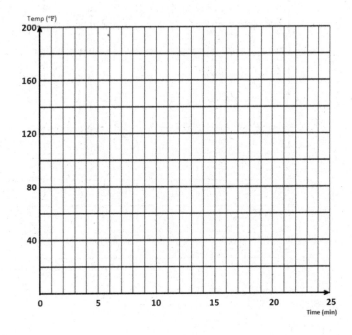

EUREKA
MATH™

c. Coffee is safe to drink when its temperature is below 140°F. How much time elapses before each cup is safe to drink?

3. Suppose Cup 2 is poured 5 minutes after Cup 1 (the pot of coffee is maintained at 180°F over the 5 minutes). Both are left sitting in the room that is 75°F.

a. Use the equation for Cup 1 found in Exercise 1, part (a) to write an equation for Cup 2.

b. Graph and label both on the same coordinate plane, and describe how to obtain the graph of Cup 2 from the graph of Cup 1.

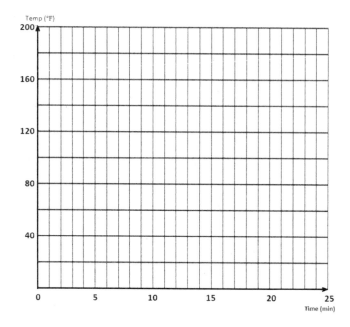

Lesson 23:   Newton's Law of Cooling

EUREKA MATH™

Problem Set

Use the Coffee Cooling demonstration on Wolfram Alpha to write a short report on the questions that follow.
http://demonstrations.wolfram.com/TheCoffeeCoolingProblem/

(Note that Wolfram's free CDF player needs to be downloaded ahead of time in order to be able to run the demonstration.)

1. If you want your coffee to become drinkable as quickly as possible, should you add cream immediately after pouring the coffee or wait? Use results from the demonstration to support your claim.

2. If you want your coffee to stay warm longer, should you add cream immediately after pouring the coffee or wait? Use results from the demonstration to support your claim.

This page intentionally left blank

Lesson 24: Piecewise and Step Functions in Context

Classwork

Opening Exercise

Here are two different parking options in the city.

1-2-3 Parking	Blue Line Parking
$6 for the first hour (or part of an hour) $5 for the second hour (or part of an hour) $4 for each hour (or part of an hour) starting with the third hour	$5 per hour up to 5 hours $4 per hour after that

The cost of a 2.75-hour stay at 1-2-3 Parking is $6 + $5 + $4 = $15. The cost of a 2.75-hour stay at Blue Line Parking is $5(2.75) = $13.75.

Which garage costs less for a 5.25-hour stay? Show your work to support your answer.

Mathematical Modeling Exercise

Helena works as a summer intern at the Albany International Airport. She is studying the parking rates and various parking options. Her department needs to raise parking revenues by 10% to help address increased operating costs. The parking rates as of 2008 are displayed below. Your class will write piecewise linear functions to model each type of rate and then use those functions to develop a plan to increase parking revenues.

Parking Rates (Effective October 28, 2008)

Short Term Rates
Located on first floor of parking garage and front of the terminal

First Half Hour:	FREE
Second Half Hour:	$2.00
Each Additional Half Hour:	$1.00
Maximum Daily Rate:	$24.00

Garage Parking Rates
Located on floors two, three, four and five of the parking garage

First Hour:	$2.00
Each Additional Hour:	$2.00
Maximum Daily Rate:	$12.00
Five Consecutive Days:	$50.00
Seven Consecutive Days:	$64.00

Long Term Parking Rates
Located behind the parking garage

First Hour:	$2.00
Each Additional Hour:	$1.00
Maximum Daily Rate:	$9.00
Five Days:	$36.00
Seven Days:	$45.00

Economy Remote Lot E - Shuttle to and from Terminal

First Hour:	$1.00
Hourly Rate:	$1.00
Maximum Daily Rate:	$5.00

1. Write a piecewise linear function using step functions that models your group's assigned parking rate. As in the Opening Exercise, assume that if the car is there for any part of the next time period, then that period is counted in full (i.e., 3.75 hours is counted as 4 hours, 3.5 days is counted as 4 days, etc.).

Helena collected all the parking tickets from one day during the summer to help her analyze ways to increase parking revenues and used that data to create the table shown below. The table displays the number of tickets turned in for each time and cost category at the four different parking lots.

Parking Tickets Collected on a Summer Day at the Albany International Airport

Short Term			Long Term			Parking Garage			Economy Remote		
Time on Ticket (hours)	Parking Cost ($)	Number of Tickets	Time on Ticket (hours)	Parking Cost ($)	Number of Tickets	Time on Ticket (hours)	Parking Cost ($)	Number of Tickets	Time on Ticket (hours)	Parking Cost ($)	Number of Tickets
0.5	0	400	1	2	8	1	2	8	1	1	
1	2	600	2	3	20	2	4	12	2	2	
1.5	3	80	3	4	24	3	6	8	3	3	
2	4	64	4	5		4	8	4	4	4	
2.5	5	8	5	6		5	10	0	5	5	
3	6	24	6	7		6	12	16	5 to 24 hrs	5	84
3.5	7	4	7	8	60	6 to 24	12	156	2 days	10	112
4	8		8	9	92	2 days	24	96	3 days	15	64
4.5	9		8 to 24	9	260	3 days	36	40	4 days	20	60
5	10		2 days	18	164	4 days	48	12	5 days	25	72
5.5	11		3 days	27	12	5-6 days	50	8	6 days	30	24
6	12		4 days	36	8	7 days	64	4	7 days	35	76
6.5	13		5 days	36	20				8 days	40	28
7	14		6 days	36	36				9 days	45	8
7.5	15		7 days	45	32				10 days	50	4
8	16	4							14 days	70	8
8.5	17								18 days	90	4
9	18	8							21 days	105	4
9.5	19										
10	20										
10.5	21		For example, there were 600 short term 1-hr tickets charged $2 each. Total revenue for that type of ticket would be $1200.								
11	22										
11.5	23										
12 to 24	24	8									

2. Compute the total revenue generated by your assigned rate using the given parking ticket data.

3. The Albany International Airport wants to increase the average daily parking revenue by 10%. Make a recommendation to management of one or more parking rates to change to increase daily parking revenue by 10%. Then, use the data Helena collected to show that revenue would increase by 10% if they implement the recommended change.

Problem Set

1. Recall the parking problem from the Opening Exercise.

    a. Write a piecewise linear function P using step functions that models the cost of parking at 1-2-3 Parking for x hours.

    b. Write a piecewise linear function B that models the cost of parking at Blue Line parking for x hours.

    c. Evaluate each function at 2.75 and 5.25 hours. Do your answers agree with the work in the Opening Exercise? If not, refine your model.

    d. Is there a time where both models have the same parking cost? Support your reasoning with graphs and/or equations.

    e. Apply your knowledge of transformations to write a new function that would represent the result of a \$2 across-the-board increase in hourly rates at 1-2-3 Parking. (Hint: Draw its graph first, and then use the graph to help you determine the step functions and domains.)

2. There was no snow on the ground when it started falling at midnight at a constant rate of 1.5 inches per hour. At 4:00 a.m., it starting falling at a constant rate of 3 inches per hour, and then from 7:00 a.m. to 9:00 a.m., snow was falling at a constant rate of 2 inches per hour. It stopped snowing at 9:00 a.m. (Note: This problem models snow falling by a constant rate during each time period. In reality, the snowfall rate might be very close to constant but is unlikely to be perfectly uniform throughout any given time period.)

    a. Write a piecewise linear function that models the depth of snow as a function of time since midnight.

    b. Create a graph of the function.

    c. When was the depth of the snow on the ground 8 inches?

    d. How deep was the snow at 9:00 a.m.?

3. If you earned up to \$113,700 in 2013 from an employer, your social security tax rate was 6.2% of your income. If you earned over \$113,700, you paid a fixed amount of \$7,049.40.

    a. Write a piecewise linear function to represent the 2013 social security taxes for incomes between \$0 and \$500,000.

    b. How much social security tax would someone who made \$50,000 owe?

    c. How much money would you have made if you paid \$4,000 in social security tax in 2013?

    d. What is the meaning of $f(150,000)$? What is the value of $f(150,000)$?

4. The function f gives the cost to ship x lb. via FedEx standard overnight rates to Zone 2 in 2013.

$$f(x) = \begin{cases} 21.50 & 0 < x \le 1 \\ 23.00 & 1 < x \le 2 \\ 24.70 & 2 < x \le 3 \\ 26.60 & 3 < x \le 4 \\ 27.05 & 4 < x \le 5 \\ 28.60 & 5 < x \le 6 \\ 29.50 & 6 < x \le 7 \\ 31.00 & 7 < x \le 8 \\ 32.25 & 8 < x \le 9 \end{cases}$$

a. How much would it cost to ship a 3 lb. package?

b. How much would it cost to ship a 7.25 lb. package?

c. What is the domain and range of f?

d. Could you use the ceiling function to write this function more concisely? Explain your reasoning.

5. Use the floor or ceiling function and your knowledge of transformations to write a piecewise linear function f whose graph is shown below.

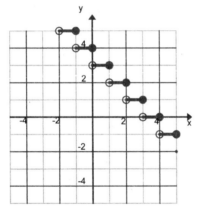

Lesson 24:   Piecewise and Step Functions in Context

EUREKA
MATH